आनन्दाम्बुधिवर्धन प्रातिपद पूर्णामृताख्व... ...आनन्द

र्वात्मस्नपनं परं विजयते श्रीकृष्णसङ्की... ...रु

नाम्नामकारि बहुधा निजसर्वशक्ति- ना... ...म

: स्त्रार्पिता नियमितः स्मरणे न कालःभगवन्ममापि स्त्रार्पिता स्त्रार्

एतादृशी तव कृपा भगवन्ममापि भगवन्ममापि नाम्नामकारि एतादृश

दुर्दैवमीदृशमिहाजनि नानुरागःतव भगवन्ममाज्ञिानन्दाम्बुधिव दुर्दैव

णादपि सुनीचेन तरोरिव सहिष्णुना सहिष्णुना नाम्नामकारि तृणादपि

ममानिना मानदेन कीर्तनीय सदा हरिः सदा आनन्दाम्बुधिव अमानि

: न धनं न जनं न सुन्दीं कामये कीर्तनीय सदा हरिः स्त्रार्पिता न ध

कवितां वा जगदीशः कामये सदाजन्मनीश्वरे आनन्दाम्बुधिव कवि

मम जन्मनि जन्मनीश्वरे मां विषमे भवाम्बुधौ    सदा हरिः मम ज

भवताद्भक्तिरहैतुकी त्वयिविषमे भवाम्बुधौभवताद्भक्तिरहैतुकी भवता

यि नन्दतनुज! किङ्करं किङ्करं अयि नन्दतनुज!मम जन्मनिअयि नन्

ततं मां विषमे भवाम्बुधौ मां विषमे भवाम्बुधौ सदा हरिःपतितं मां

कृपया तव पादपङ्ज- अयि अयि विचिन्तय आनन्दाम्बधिव कृप

दाम्बुधिवर्धन प्रातिपद् पूर्णामृतास्वादन आनन्दाम्बुधिव आनन्दाम्बु

स्नपनं परं विजयते श्रीकृष्णसङ्कीर्तनम् सर्वात्मस्नपनं सर्वात्मस्नपनं

मकारि बहुधा निजसर्वशक्ति- नाम्नामकारि नाम्नामकारि नाम्नामकारि

र्पिता नियमितः स्मरणे न कालःभगवन्ममापि स्त्रार्पिता स्त्रार्पिता नि

शी तव कृपा भगवन्ममापि भगवन्ममापि नाम्नामकारि एताद्दशी त

मीद्दशमिहाजनि नानुरागःतव भगवन्ममाज्ञिआनन्दाम्बुधिव दुर्दैवमीद्दशा

द्रपि सुनीचेन तरोरिव सहिष्णुना सहिष्णुना नाम्नामकारि तृणादपि सु

निना मानदेन कीर्तनीय सदा हरिः सदा आनन्दाम्बुधिव अमानिना म

नं न जनं न सुन्दीं कामये कीर्तनीय सदा हरिः स्त्रार्पिता न धनं न

तां वा जगदीशः कामये सदा जन्मनीश्वरे आनन्दाम्बुधिव कवितां वा

जन्मनि जन्मनीश्वरे मां विषमे भवाम्बुधौ      सदा हरिः मम जन्मनि

क्तिरहैतुकी त्वयिविषमे भवाम्बुधौैवताद्भक्तिरहैतुकी भवताद्भक्ति

नन्दतनुज! किङ्करं किङ्करं अयि नन्दतनुज!मम जन्मनि अयि नन्दत

मां विषमे भवाम्बुधौ मां विषमे भवाम्बुधौ सदा हरिःपतितं मां वि

तव पादपङ्ज- अयि अयि विचिन्त्य आनन्दाम्बुधिव कृपया तव

YOUR HEALTH, YOUR DESTINY, YOUR CHOICE

# WELLNESS AT
# WARP
# SPEED

## DR. NOAH McKAY M.D.

MANDALA

# MANDALA
PUBLISHING

Mandala Publishing
17 Paul Drive
San Rafael, CA 94903
www.mandala.org
415.526.1370
Fax: 415.526.1394

Orion Research Institute, Inc.
P.O. Box 886
McMinnville, OR 97128
www.WellnessatWarpSpeed.com
888.899.1149
Fax: 267.295.2585

Library of Congress Cataloging-in-Publication Data available.

ISBN-13: 978-1-601091-08-6

Palace Press International, in association with Global ReLeaf, will plant two trees for each tree used in the manufacturing of this book. Global ReLeaf is an international campaign by American Forests, the nation's oldest nonprofit conservation organization and a world leader in planting trees for environmental restoration.

10 9 8 7 6 5 4 3 2 1

Printed in China by Palace Press International

## DEDICATION

I dedicate this book to all men
and women who willingly embrace
discomfort and adversity and test the
limits of their physical and spiritual
powers in pursuit of the truth. To
this growing multitude of noble souls,
illness, disability, divorce, bankruptcy
and prison are opportunities in
disguise, chances to reconnect with
higher selves as a part of the infinite
source of healing that unifies and
connects us all. With God's blessing
and the support of loved ones, let's
bring Warp Speed Peace, Happiness
and Wellness to our families, our
communities and our world.

# CONTENTS

# ARE YOU READY FOR WARP SPEED WELLNESS?

**warp speed:** (ˈworp spēd) n. faster than light velocity achieved by bending space/time to reduce the distance between two points

**wellness:** (ˈwel-nəs) n. good health

A revolution is under way, one that will soon touch every aspect of your life. It began quietly and without fanfare nearly a century ago in the minds of a few keen observers who looked at the world through the eyes of imagination and saw something more—something utterly impossible and yet unquestionably true. What these men discovered is about to change the way you think, the way you live, and perhaps most importantly, the way you heal.

It has touched my life already. The changes came as unexpectedly as an earthquake and unrelenting series of aftershocks: life-threatening illness, miraculous recovery, astonishing success in business and medicine, incarceration in a federal prison, open-heart surgery—and more miracles.

I am a physician, schooled in the world's finest medical traditions, but in the fall of 1989 all my years of medical training, practice and theory came crashing down around me when I was suddenly stricken by an incurable, incapacitating and near-universally fatal medical condition called viral cardiomyopathy. The prognosis was grim. The disease is especially deadly in young people; at age thirty-three, my life expectancy dropped to just twenty-four months. I became an invalid overnight.

Confined to my bed and utterly dependent on oxygen, there was little to do but read—just as you are reading now. That's where the miracles began.

I am here today to tell you my story because what I learned from the work of an Irish physicist by the name of John Stewart Bell saved my life—and it can save yours too.

Bell's Theorem was my introduction to the strange new world of quantum physics, where the routine assumptions of everyday life no longer apply. The implications of Bell's work struck with the full force of an epiphany. In that instant, I suddenly understood many of medicine's most challenging puzzles in a new

light. I was granted passage to the realm of miracles. I found hope.

*In 1964, John Stewart Bell offered mathematical proof validating the existence of non-local interactions. Bell's Theorem was experimentally verified by a team of French physicists led by Alain Aspect in 1982.*

*Wellness at Warp Speed*

I found my path to warp speed wellness.

Are you ready to find yours?

Everything in the universe—including your body—is made of atoms. The atoms and sub-atomic particles that compose your body's ten trillion cells are constantly in motion, traveling at near the speed of light—186,000 miles (300,000 kilometers) per second!

If our bodies are operating at light speed on the inside, at our most fundamental level of existence, why don't we feel it on the outside, where we spend our days living at the painfully slow pace of ordinary mortal beings? Why is there a gap between these two worlds?

*Plato (428? – 347 BC)*

### The invisible is greater than the visible.

– Plato

We don't feel the speed of cellular operations within our bodies for the same reason we don't feel the speed of the Earth moving through space. At this very moment, the planet beneath your feet is rocketing around the sun at an astonishing 66,780 miles per hour—and yet you don't feel it. Faster still is the speed of our solar system, racing through the galaxy at a staggering 487,353 miles per hour! You don't feel that either. Why?

Your brain and the neural network it controls compose the most sophisticated data management system in the world, one that is constantly working faster than all of the fastest supercomputers at the US Pentagon combined. Yet even this remarkable processing system would have a difficult time coping

*Albert Einstein struggled with the quantum notion of "spooky action at a distance," known today as non-locality.*

Ferdinand Schmutzer

with the tremendous influx of data reaching you every second. Your body has been brilliantly designed to shelter you from the risk of sensory overload by the positioning of five sense organs between you and the outside world.

Our five primary senses—sight, smell, hearing, taste and touch—are often called the gateways of perception. It is a good description. We receive information through our senses, but like all effective gateways, they are narrow and don't allow everything to enter at once. The job of your eyes, ears, nose, tongue, and skin is to organize and translate the massive quantity of information you receive from the outside world into manageable bits of data.

Without a reliable way to control the barrage of incoming data, you would be constantly overwhelmed and exhausted. Your brain is already too busy processing its more than fifty thousand thoughts a day to be bothered with a continuous stream of information about the earth's rotational speed and trajectory and the movement of subatomic particles inside your body.

While most of us limit our awareness to events taking place at the relatively slow pace of everyday life, it is important to remember that what we perceive through our primary senses represents less than one tenth of 1% of what is actually going on around us—or within us. It's the rest—that which lies just beyond the reach of ordinary perception—that makes the biggest difference in our lives.

The Greek philosopher Plato was right when he observed in 440 BC, "The invisible is greater than the visible." What history's most inspiring spiritual leaders have always understood is that our survival and evolution as individuals and as a species are dependent on how quickly we broaden our bandwidth of awareness and understanding—because it is in the realm of the invisible that we humans have the greatest impact on our own destiny and wellness.

But how can we ordinary humans reach beyond the limits of our own sensory pathways?

A thousand years ago, the answer to that question would have been available only to a handful of initiates devoted to the full-time pursuit of the spiritual life. Not any more. If you are reading these words, congratulate yourself. You are living in the age of the greatest revolution in the history of mankind—the quantum revolution.

When marking the greatest achievements of the twentieth century, technological advances like space travel, the development of antibiotics and the invention of the computer often top the list. History will have a very different view of our times. Two hundred years from now, our great-grandchildren will honor the twentieth and twenty-first centuries for the remarkable unifying discoveries they fostered in the field of quantum physics.

By the early years of the twentieth century, physicists had developed two different systems to describe the universe. Astrophysicists exploring the vast reaches of the cosmos established one set of rules to explain events in the world of the very large—planets, stars, and galaxies—but an entirely different model to describe the bizarre world of the very

small: atoms and subatomic particles. Quantum physicists took up the challenge and began mapping the tiny world of the atom, where impossible events are the norm and rules of common sense no longer apply.

Pioneering physicists exploring in the field of quantum science were shocked by the implications of their own discoveries. Even Albert Einstein, renowned for his genius and "out of the box" thinking, found it difficult to reach beyond the confines of human logic and his background in Newtonian science to embrace the absurdly strange new field of quantum physics.

Einstein was particularly troubled by the notion that paired quantum particles could be linked in such a way that measurements made on one particle would simultaneously affect the other regardless of the distance between them—a concept known as **non-locality**. If communications were somehow passing between the separated particles, then that communication was occurring faster than the speed of light. Einstein believed this to be impossible.

Dismissing the notion of such intimate connections as "spooky action at a distance," in 1935 he and colleagues Boris Podolsky and Nathan Rosen engaged in a well-publicized series of debates with Danish physicist Niels Bohr about what came to be known as the EPR Paradox.

Most physicists weighing in on the debate sided with Bohr, but nearly three decades passed before John Stewart Bell stepped forward with a mathematical proof proposing an experimental means of settling the question once and for all. In 1982, a team of French physicists led by Alain Aspect finally carried out Bell's proposed experiment.

The results were conclusive and irrefutable: Einstein had been wrong.

Light speed isn't the end—it's just the beginning.

Non-locality rules the quantum universe, proving physicist David Bohm's assertion that we live in an undivided and indivisible universe, one that is unified and whole no matter how hard we try to partition or fragment it. It is the

**Every known disease has a known cure!**

same essential truth encapsulated in the core teachings of Krishna, Zoroaster, Buddha, Jesus, Mohammad and the Dalai Lama. Non-locality provides us with solid mathematical proof validating the spiritual teachings humanity has long struggled to put into practice. What you do today matters. Your actions and decisions affect everything and everyone around you regardless of where they are. We are all connected. There is only one nation, one family, one God to serve and behold. The rest is semantics.

Beyond its broad philosophical implications, the new science of quantum physics also provides a useful owner's manual for those of us living with the delights, mysteries and occasional malfunctions of our high-speed bodies.

Stunning advances in the field of high-speed medical diagnostics such as CT, PET and MRI scanners are all byproducts of the quantum revolution in health care—but these first generation devices offer only a glimpse of health care's amazing future. Quantum technologies exist today, often below the public radar, that allow for not only rapid diagnosis but quick, painless treatment of seemingly incurable diseases.

Which brings us to a little insider secret: every known disease has a known cure. Yes, you read that right—*Every Known Disease Has a Known Cure!* This is not something widely taught in Western medical schools, but it is true nonetheless. If you've been diagnosed with an "incurable" disease, get a second opinion—and a third and a fourth if necessary. Don't be afraid to broaden your search beyond the borders of your home country. Solutions exist, and you will find them in surprising places.

Caring for thousands of patients during my twenty-year career as a general practitioner taught me a great deal about healing. I have tremendous respect for the high-tech tools of my trade and advances made in the field of pharmacology—but the most remarkable discovery of my career had little to do with new pharmaceutical breakthroughs or technological advancements. It had everything to do with the remarkable natural healing abilities of each patient. Those patients who successfully unlocked their body's quantum healing energy recovered in record time, while others with the same prognosis died. You are in charge of

*Wellness at Warp Speed*

your own destiny. Miraculous recoveries occur in those who adopt an attitude of self-awareness and self-determination and the intrepid outlook of an explorer.

Join us as we embark on a voyage of discovery to the mystical reaches of your inner universe.

Whatever your health care challenges, bring them along. You will find hope, courage and inspiration every step of the way.

**PHYSICIAN, HEAL THYSELF**

How would you define a perfect life: Radiant health? A rewarding career? Good friends and a loving family?

If you have all of them, congratulations. You have much to be grateful for. Happiness and optimism come easily when things are going well in our lives, and a positive outlook is a vital component of any wellness program.

However, if you are old enough to read this book, you've probably already noticed that life doesn't always run smoothly. Brakes fail, wars erupt, relationships crumble, pink slips unexpectedly turn up in the inbox. We dread hearing words like "biopsy" because a few simple letters have the power to transform our most terrifying nightmares into stark waking reality.

What then?

What happens when the crisis hits home?

## SHATTERING ILLUSIONS

If your life is less than perfect, relax, take a deep breath—and let me be the first to welcome you aboard! However unlikely it may seem, you have just embarked on what you may one day recognize as the most rewarding adventure of your life. I know. I've been there.

I am a doctor. Medicine has been my passion and calling since childhood. Five years into my career—just as the decades of hard work and study required to become a doctor had begun to pay off—everything changed.

The morning of September 21, 1989 began much like any other, with the familiar tone of my Casio alarm clock signaling the start of another busy day. But this day, something was different. Very different. When I reached out to silence the noise, nothing happened. I couldn't move a muscle. The effort to move my arm left me gasping for breath, and the sound of the alarm was slowly drowned out by a deafeningly loud thumping sound coming from inside my chest.

I lay frozen in bed, utterly helpless, my mind racing in a thousand directions at once. Something was seriously wrong—but what?

The only thing clear to me about my situation was the fact that I was alone and in desperate need of help. With a herculean act of will, I reached for the phone and arranged an emergency visit with a cardiologist.

The usual gauntlet of testing began with my arrival at the hospital. A healthy, active thirty-three-year-old physician became an invalid overnight, and no one had any idea why.

## THE WORLD AT DEATH'S DOOR

I awoke in the wee hours of the morning in the throes of a second catastrophic collapse. I was rushed to the hospital and settled into a room with the usual intravenous treatment regimen for patients in heart failure. My oxygen flow meter was set at three liters per minute—about 50% higher than most cardiac patients receive—but I wasn't about to complain. I was grateful for every breath.

The abruptness of it all bothered me, the fact that the collapse had come without warning. I replayed the events of the past fourteen days over and over again in a futile search for answers until my wife finally intervened with calm, methodical logic: "You're overthinking. You should conserve your energy and use it to focus on healing."

Kim's reassuring presence brought me great comfort. I secretly wanted her to stay and keep me company, but the tiny, barren room didn't even offer an armchair. The last thing I remember was urging her to go home as I drifted off to sleep.

I awoke ninety minutes later, alone, terrified and gasping for breath. I pressed the call button again and again for help, but the nurses didn't respond. Groggy, shaky and nearing a state of panic, I fumbled with the oxygen regulator valve myself and somehow managed to double the flow before collapsing back onto the bed.

When the nurse finally arrived, she immediately set into play a terrifying farce that would nearly kill me. Although I was in crisis and acutely short of breath, she walked to the wrong side of the bed—away from the equipment—and stood there patiently awaiting orders I was in no shape to give. She was clearly intimidated by the fact that I was a doctor. "What do you want me to do?" she asked.

"PLEASE take my blood pressure," I gasped, barely able to breathe.

My blood pressure had plummeted; it took her three tries to get a reading. When she finally realized how close I was to death, she took action. Before I could say another word, she'd raised the top of my bed and stuffed a second pillow beneath my head.

Her misguided efforts slammed me into another dimension. I suddenly found myself surrounded by hundreds, perhaps thousands of heart patients, each dying from hypoxia—lack of oxygen. I was right there with them—all of them—simultaneously, feeling their pain and hearing their muted screams. They had come to give me a message. They wanted me to understand their suffering—really understand it—and now I did. I knew exactly how and why they'd died.

The epiphany faded into the faint image of a nurse fiddling with an oxygen dial on the wall. In my final conscious effort, I grabbed her by the wrist, pointed to my feet and gasped, "Legs up. Legs *up*."

And then I was gone.

Recognizing her mistake, the nurse rushed to undo it by raising my feet and lowering my head. As a life-sustaining flow of blood slowly resumed in my brain, desperately needed oxygen began seeping into my cells one by one, calling me back from the brink of death.

When the haze finally lifted, my first thought was one of escape. An incompetent nurse had nearly killed me just hours after my arrival at the hospital. The facility was clearly unprepared to care for someone in my condition, so I dismissed the cardiologist's suggestion of open-heart surgery and arranged my own discharge in the morning.

The night's strange events gave me a lot to think about. My encounter with the dead had demonstrated the presence of invisible and ephemeral connections that exist beyond the reach of logic, medical training or common sense. I now understood that given the right conditions, the mind has the capacity to access nonpersonal history and information and play it back frame by frame for analysis.

The human heart is much more than a mechanical pumping device with four chambers, four valves and a handful of connecting vessels. Our bodies are intricately intertwined with the vast fabric of the universe. We access that connection through the heart.

I began my third day flat on my back staring at the bare white ceiling of the hospital's catheterization lab, waiting to go another round with the cardiologist. It's funny what runs through your mind at a time like that. My thoughts drifted idly from one grim possibility to the next until I suddenly noticed that the entire room had been tiled from floor to ceiling in white. It bothered me that nothing about the place suggested healing in progress. I'd been in hundreds of rooms like it before, but I'd never seen one from this distorted, upside-down angle. I wondered if the contractor had been paid extra for making the place look like a prison.

And then it was time. The squeal of a swinging door announcing the cardiologist's arrival sent an involuntary shiver rippling across my body. Someone was about to send a large-bore catheter snaking its way from my groin into my heart. Memories of cardiac catheterization procedures I'd performed myself hovered in the room like ghosts, and I suddenly felt deeply embarrassed by my own lack of sensitivity to the fears of my patients. I swore that if I survived the ordeal intact, things would be very different.

My cardiologist was one of the most caring individuals I've encountered in my profession, but he had trained—as I had—in the Western tradition of medicine. We are taught to be competent technicians, but we are seldom reminded that we are working on human beings. Now that I was the patient, I understood things differently. I desperately needed someone to walk through the door, hold my hand and promise me that all would be well. No one came.

The verdict arrived the next day: my heart was under attack. This wasn't a classic heart attack triggered by blockages in the coronary arteries—it was a severe, life-threatening invasion of my heart by an unknown viral intruder. The infection had caused extensive damage to my heart muscles and rendered them too weak to function.

I was in acute heart failure. The cardiologist and I both knew that the chances of full recovery in a thirty-three-year old male diagnosed with acute viral cardiomyopathy are extremely low. There was little more to be said.

For the first time in my life, I felt totally helpless and vulnerable. Modern cardiology gave me the best cardiac drugs it had to offer, but none of them could change the fact that I was now a terminal patient. I would spend the rest of my life confined to bed by the disease that would soon kill me.

I wanted things to be different. I wanted someone to tell me that I would be fine, that it had all been a terrible mistake.

I felt pity for myself.

> Hope is invisible and intangible, but like the forces of gravity and electromagnetism, it has a profound influence on events in the physical world.

I felt profoundly disappointed with my profession.

## INSTINCT AND SURVIVAL

The mysterious episode had changed none of the unrelenting facts of my situation. I was dying. Until I did so, I would remain confined to my bed, wholly dependent on others for my care and on an oxygen tank for my day-to-day survival.

What had changed was my perspective. My strange experience convinced me that there are regions of awareness and knowledge that exist beyond the confines of textbook medicine. Since traditional Western medicine didn't even know these realms existed, it was clear that I'd have to look elsewhere for a map of the terrain that would lead to my recovery.

My search began with two long-neglected stacks of books at my bedside: one reflected my keen interest in quantum physics, string theory and the holographic universe; the other promised a better understanding of the many faces of mysticism. Christians, Jews, Muslims, Buddhists and Hindus each held a distinct mystic tradition. I had little interest in what divided them; what I really wanted to explore were the essential truths they held in common. What could I learn by looking at mankind's spiritual legacy as a whole?

Most importantly, what could science and spirituality teach us about healing?

Days melted into weeks and weeks into months with no visible sign of recovery. Oxygen and bed rest allowed me to survive that bleak and difficult time—but a third, more powerful healing force was also beginning to make its debut in my life: HOPE.

The long hours of the night were difficult and depressing. Cardiac patients succumb to low cardiac output in the early hours of the day, usually between 1:00 and 3:00 am, and become acutely agitated and short of breath. I spent most of my nights in a state of fear and trepidation, praying that I wouldn't go into complete cardiac failure.

Every morning brought a new opportunity to experience hope for the very first time. I embraced each new day with open arms, knowing that I had survived another night and that I would one day be fine. I was alive, and for me, that was the promise of hope.

Hope is invisible and intangible, but like the forces of gravity and electromagnetism, it has a profound influence on events in the physical world. No matter how many options failed, hope remained the driving force sustaining me in the face of impossible odds. The discoveries I made in books on mysticism and quantum physics stood in stark contrast to the bleak pronouncements of the Merck manual and Harrison's textbook of internal medicine.

The bibles of medicine told me I was dying. Quantum science and mankind's great spiritual teachers told me Merck just might be wrong.

Who was I to believe?

*"Breath of Fire," Lake Washington, 1990*

## MIRACLE IN MAZATLÁN

My life was coming to an abrupt end at a most unfortunate and inopportune time. I was thirty-three, newly married, in the fifth year of a thriving medical practice and at the top of my game as a physician and aspiring businessman. This was not a good time to die! The thought of lying in bed helplessly sucking on oxygen while my hopes and dreams faded away was too much to bear. Above all, I didn't want to leave my beautiful young bride alone to cope with a heavy load of outstanding business debt.

After three months of bed-ridden disability, time was running out. Something had to be done, and done quickly.

"Let's go to Mexico," I announced late one evening when my wife arrived home from work.

"OK," she replied, as if this were the most ordinary thing in the world.

A series of miracles fell into place, and we were on a plane for Mazatlán within the week.

Family, friends and colleagues were horrified by my decision. By traveling to Mexico, I would be depriving myself of twenty-four-hour access to the life-saving 911 system and the safety net offered by modern cardiology. Worse, flying Air Mexicana would mean abandoning my oxygen tank. I might not even survive the flight.

Everyone thought I'd lost my mind, and frankly, I half agreed with them. It was a completely irrational decision, a choice made between scientific reason and pure animal instinct. In the end, instinct won. Above all, I craved warmth, sunlight and a final chance to live—or die—on a beautiful beach in paradise.

When our plane arrived in Mexico, I paused for a moment on the stairs to catch my breath before beginning the long, slow descent to the tarmac. A bright orange sunset wrapped itself around the modest silhouette of the two-story stucco terminal building to my right. I closed my eyes to savor the image and felt all my worries about oxygen tanks and health care services drifting away on my first delicious breath of warm salt air. My whole body tingled with the magical glow of late-afternoon sunlight. "I will be healed here," I said. I was sure of it.

Our first week passed uneventfully, much of it spent resting in the hotel room. My initial hunch about Mexico was correct: my body was definitely functioning better in my new location. At the end of seven days, my wife felt confident that I had improved enough to re-main in Mexico on my own while she returned to her job in Seattle.

The following day I decided to venture out and do a little exploring. Curious about a flyer I'd seen advertising a yoga workshop, I slipped into a metaphysical bookstore and made my way past aisles of incense, candles and crystals to a small yoga studio in the back.

I'd hoped to discuss my fragile condition with the instructor before class began, but it was not to be. The yogi—complete with long beard and turban—sat in full lotus position on a sheepskin rug at the far end of the room, deep in meditation. When he'd finished, he gently opened his eyes and peered out at me, bringing his hands together in the customary manner of greeting. He gestured for me to take a seat among the others, and then began to speak.

Guru Dev had exquisite timing. "You don't choose yoga—yoga chooses you," he began, pausing just long enough to let the point sink in before sharing his own remarkable story. He had contracted an incurable disease and had abandoned all hope of recovery until Yogi Bajan introduced him to the power of yogic breath. The physical and emotional transformation he experienced was so powerful that he abandoned everything to become a full-time yoga instructor.

The guru went on to explain that Kundalini

is the yoga of transformation. The Sanskrit word *yoga* means union; the meditation and exercises unify the practitioner's mind, body and spirit into a single cohesive whole. He could have talked for hours about yoga's origin in ancient India or discussed its many rich and varied traditions of practice but decided instead to let yoga speak for itself through a quick demonstration of the Breath of Fire.

He instructed us to raise our arms straight up over our heads. With elbows locked and palms pressed together as if in prayer, we closed our eyes and began inhaling and exhaling through our nostrils, filling the chest and abdomen to capacity with each breath. After ten slow, deep repetitions, we increased the pace and intensity of each breath until we sounded like an advancing army of steam engines.

In-out, in-out, in-out he pressed, pushing us to the limits of our bodies and beyond. My heart was pounding in my throat, but I stayed with it. By the time he signaled the final breath with an unusually loud, deep inhalation, I was ready to collapse.

He asked us to lie on our backs and continue with our long, deep breathing. I felt my flesh and bones melt into the carpet and into the floor and subfloor beneath. My body, soul and spirit were indeed united; I felt tingly and electric everywhere.

I was shocked and overjoyed at my body's amazing performance, but there was little time to revel in my success. It seemed like we barely hit the floor before the guru's commanding voice rang out, "OK, enough already. Rise and get ready for Breath of Fire, round two."

In the course of our two-hour Kundalini session, I worked every muscle, nerve and alveolar (air) sac in my chest. By the time the class ended, I felt more aroused and energized than I had in months. I joined Guru Dev for lunch that day, eager to learn what he would suggest for a man in my condition.

He listened patiently to my story, then shook his head and replied thoughtfully, "Doc, you are one of the dumbest doctors I've ever met."

He didn't wait for my response.

"You've based your entire future on the

opinion of just one man."

I didn't know where he was going with this, but I was too stunned to interrupt.

"You've told me what your doctor thinks you're suffering from, but I want to know what you think."

"Well," I began, "According to Harrison's and the Merck manual, the chances of survival for a thirty-three-year old male with viral cardiomyopathy are very slim."

He interrupted before I could go on, "Doc, don't tell me what the books say. They can say whatever they want to say."

In a final dramatic gesture, he sipped the last of his carrot juice and said—only half in jest—"It's clear *you're* going to need a lot of work."

Motivated by his words and by the stunning improvement I'd experienced in the first session, I practiced the yogic breath techniques for four hours the following day. My health continued to improve. I'd managed well on my own all week and could walk an entire block without becoming short of breath. By late afternoon I felt ready to test the limits of my newfound body. I was ready for the ocean.

I made my way to the beach and walked gently into the surf, bouncing on tiptoes with each advancing wave. My whole body tingled with anticipation. When the water reached my chest, I touched the seabed one last time and then released my tentative hold on the earth. I was swimming—ten yards, twenty, thirty, sixty. When I'd finally had enough, I rolled over onto my back and lazed in the surf, basking in the simple delight of warm water and sunlight caressing my skin.

I left the water short of breath but exhilarated by my unexpected success. I rested on the glimmering sunbaked sands, reveling in the world of possibilities that now lay before me. I was certain that the Breath of Fire technique would have me seeing patients in no time. I wanted to scream with joy and excitement but saved my enthusiasm for a call to my wife later that evening.

"You won't believe what I did today."

"What?" Kim asked tentatively, never quite sure where a conversation with me might lead.

"I went for a swim. In the ocean. Alone

*"Hope,"* a self-portrait painted in January 1990, the fourth month of my convalescence period

and without endangering myself or the lifeguard."

She broke out into a laugh. "I'm proud of you, sweetie, but don't push it."

"Why don't you call the office and tell them to put me back on the schedule at work?" I urged.

"I think you should come home first, honey," she stalled. "We can discuss it when you get here."

Despite my wife's skepticism, I knew in my heart that I could be ready. A cure had materialized before my very eyes, and it had come in a most unexpected way. A guru on a sunny beach in Mexico had given me a personal tour of a world my profession didn't even know existed.

Nothing about my experience seemed either plausible or rational—but it was undeniably real. In two weeks I'd been transformed from a depressed, bed-ridden cardiac cripple making end-of-life plans into a vibrant new man racing through life with the electrifying excitement of a runner nearing the finish line!

As a physician, I could no longer dismiss accounts of spontaneous recovery in terminal patients as unscientific or anecdotal. The unbelievable physiologic transformation I experienced was all the proof I needed to know that miracles exist. Spontaneous healing happens, and it can be triggered even in skeptical physicians like me.

It was clear that my colleagues and I still had a long way to go before we would fully unravel the deeper mysteries of the human body. I couldn't explain the cellular mechanisms or science behind what I'd experienced, but I was determined to figure it out. I wanted the entire world to know how our darkest hours of disease and despair can be transformed in a single day!

## TRANSFORMATION

I returned to my career six months after I fell ill, determined to leave no stone unturned in my exploration of promising non-traditional medical strategies. I visited acupuncturists, herbalists, homeopaths, massage therapists, chiropractic and naturopathic doctors—and even psychics and hypnotherapists. I switched to organic food, began juicing fruits and vegetables, practiced Kundalini yoga, prayed, meditated and took an intensive course with a visiting Qi-Gong master from mainland China. I took up portrait and landscape painting and filled my days with music.

I found my examination of the different traditions of health care intriguing and useful, but I wasn't ready to trade my Western medical education for any of them just yet. What I really wanted to know was why *all* these techniques stood the test of time. How could such radically different systems of care all enjoy the success they did?

I wanted to find the common thread that bound them together. I sensed a single hidden principle at work behind practices as diverse as acupuncture and anesthesia, Reiki and radiation oncology, homeopathy and endocrinology, and I was determined to find it. The search for a single unifying principle of healing

As a medical student I was told it would be wrong to give my patients 'a false sense of hope.' Today I know better. Hope is the first step in the recovery process, and no one is better positioned to take advantage of its powerful healing potential than you. There is nothing 'false' about having the imagination and courage to begin manifesting a healthy future that is not yet visible to others. Don't be afraid to lose yourself in the ecstasy of the moment and envision your preferred future as one of peace, health and happiness.

*Mahatma Gandhi (1869 – 1948)*

became my quest for the holy grail of medicine.

The following year I attended a physicians' seminar featuring a presentation on *The Non-pharmacologic Treatment of Chronic Disease* by Dr. Deepak Chopra, a doctor whose recently released book *Quantum Healing[1]* was just

---

[1]Chopra, Deepak, *Quantum Healing* (New York, Bantam Books, 1989)

beginning its climb up the New York Times Bestseller List. I had long been convinced that existing scientific models could explain the miraculous healing I experienced on the beach in Mazatlán. Dr. Chopra's talk convinced me that my beliefs about this new dimension of healing were right on target. No matter how unconventional my thoughts might seem to the more conservative members of my profession, at least one colleague agreed with me.

Inspired by this affirmation, I continued my quest as both an intellectual and experiential pursuit. The former shed metaphorical light on my understanding through the works of the giants of twentieth- century physics—Bohr, Einstein, Schrödinger, Heisenberg, Bohm and Bell—who collectively unraveled the mysteries of the atom and provided solid mathematical models for understanding the fundamental basis of reality. The latter would provide a very different sort of enlightenment.

## DOING THE IMPOSSIBLE

I'd long been fascinated by the seemingly impossible feats of yogis and saints described in Swami Paramahansa Yogananda's book *Autobiography of a Yogi*.[2] I was particularly intrigued by individuals who abandoned the practice of eating and learned to sustain themselves on air, sunlight and prayer.

In 1993, I jumped at the chance to attend a two-day seminar hosted by a well-known breatharian. The presentation was neither inspirational nor scientific, and I learned nothing about how to become a breatharian. I left the seminar annoyed at having wasted $300 and an entire weekend.

As I considered the possibility of asking for my money back, I was struck by a revolutionary idea: *Why not stop eating and*

> As human beings, our greatness lies not so much in being able to remake the world — that is the myth of the atomic age — as in being able to remake ourselves.
>
> – Mahatma Gandhi

*see what happens?* Without another moment's thought, I embarked on one of the greatest adventures of my entire life.

I set no specific goals or targets — it just happened. The next morning I skipped breakfast. Later in the day I declined a lunch offer from a friendly pharmaceutical rep. I continued like this for three days, skipping one meal at a time, never thinking ahead or looking back.

By the third day, my stomach growled loudly and painfully all day. I was completely miserable and convinced that I had come to the end of my rope.

Much to my surprise, I awoke the following morning to an entirely new plateau of experiences. My energy level doubled overnight. I felt light and exuberant, and I was no longer hungry. I felt like Chuck Yeager breaking the sound barrier. My body operated at Mach I, with seemingly unlimited stores of energy, hope and passion.

---

[2]Yogananda, Paramahansa, *Autobiography of a Yogi* (Los Angeles, Self-Realization Fellowship, 1946)

victory celebration by drinking one ounce of goat milk in tribute to Mahatma Gandhi's traditional method of fasting.

I continued taking nothing but one ounce of goat milk every third day for another three weeks. I lost a pound a day throughout the fast; by the twenty-ninth day I was gaunt and thin, but I was soaring with energy and inspiration.

On the thirtieth day, I woke up light-headed, disoriented and utterly exhausted. My pulse was faint and barely palpable. I was dying.

Not wanting to alarm my sleeping wife with a situation neither of us could change, I left her undisturbed and headed for the hospital alone.

When I reached the turn-off to the hospital, the car seemed to have a mind of its own. It bypassed the exit and continued down the freeway toward the clinic. A heavy downpour made driving on the busy freeway a night-mare. I was terrified of dying

I was sick and tired of being dependent on prescription medications, so on the seventh day of the fast I discontinued the heart medications I'd been taking ever since the onset of my illness. Stopping cardiac medications abruptly usually leads to acute heart failure. My wife and I stayed up most of the night worrying that this would be my fate—but it was not. My heart didn't fail, and I awoke the next morning feeling invincible. I indulged in a small

…a thousand bolts of lightning struck the base of my spine, shot up my spinal column and exited through the crown of my head.

behind the wheel and causing an accident.

I tried to calm myself and regain a sense of control by chanting my favorite mantras, but nothing changed. I began an open dialog with God, begging for reassurance that something greater than myself existed. Death was imminent, and I desperately needed answers. I needed to know that my life had not been without purpose and meaning. I needed assurance that the miraculous events of the past thirty days had not been a product of my imagination.

When no answer came, my pleas grew louder and angrier. I demanded a sign—any sign. The event that followed marked the greatest turning point of my life. Before I'd even finished speaking, a thousand bolts of lightning struck the base of my spine, shot up my spinal column and exited through the crown of my head. For a few indescribable moments, I existed as pure light, pure energy, fully conscious and fully present. I had instantaneous and simultaneous access to all the knowledge of the universe and felt infinite energy pouring through my being.

I hadn't lost consciousness—I had entered a vastly expanded state of *awareness*. When I returned to a more "normal" state of perception, I discovered that the blast of energy had thrown my glasses to the floor and left me literally driving blind. I wept with joy all the way to the clinic. When I walked in the door, it was immediately apparent that the change I had undergone was apparent to others: a wide-eyed office manager took one look at me and spirited me away to an exam room to ask what had happened.

I didn't have the words to describe it to her then; what I experienced would later be explained to me as a full Kundalini awakening.

Whether I could explain it then or not, the experience marked the most significant turning point of my life and left me a changed man. I accessed an energy source that accomplished in seconds what traditional medicine had been unable to achieve for my patients in years. My days as a traditional prescription-pushing physician were over. I'd been transformed: I had become a quantum healer.

I resumed eating the following day. Going

without food—or water—for thirty days taught me important lessons about the process of extended fasting. I've never felt the need to engage in another long fast, but if I were to try it again, I would do things very differently. Extended fasting is not something to be taken lightly. It is definitely not something I recommend to others.

## A NEW KIND OF DOCTOR

I have known since my earliest days that I would one day be a doctor. My passion for healing has been a constant guide on a voyage of discovery few could have imagined. Who could have foreseen that my early fascination with the mechanics of the heart would one day play out in my own catastrophic illness and miraculous recovery?

My search for an understanding of the dramatic events unfolding in my life was a long one, and it led me to the discovery of surprisingly effective healing modalities unfamiliar to most in my profession.

I knew from firsthand experience that many of the therapies overlooked by my profession were real, they were available, and they worked. If they had been effective in my own life, how could I withhold them from my patients?

My career began in 1985 when I joined a team of physicians practicing at an urgent-care clinic in Washington state. I loved every minute of it—the challenge of diagnosis, the joy of watching patients recover and the indescribable satisfaction of knowing I made a difference in their lives when they did.

Two years after joining the practice, I accepted the offer of a partnership with the growing chain and opened a new branch location of my own. Much to my delight, the practice flourished.

Twenty-four months later my career was swept away by a devastating attack of viral cardiomyopathy. Against all odds, I not only survived the attack but returned to work hale, hardy and eager to share the exciting new discoveries that had led to my own healing.

I taught my patients the Breath of Fire and openly discussed alternative treatment strategies with those who were interested. Most patients were receptive to new ideas, but some physicians in the practice were not. One by one, each of the senior staff took me aside to remind me that we were running a conservative medical practice. The Breath of Fire might be an effective healing tool, but it was embarrassing to have people breathing out loud right there in the medical clinic. Couldn't I just stick with a more conventional approach?

I heeded their advice, but only for a week. My conscience nagged at me the entire time. I knew in my heart that it was time for a new kind of medicine. I also knew that the old guard would never let it happen. I did the only thing I could: in 1993, I bought the practice.

The 1990s were a time of turbulence and upheaval on two health care fronts. The entire nation had gone lawsuit-happy; this resulted in soaring insurance premiums for businesses, home owners and professionals in every field. Health care providers were especially hard hit by dramatic increases in malpractice premiums. This led to the practice of defensive medicine—an overreliance on expensive, high-tech testing among doctors who suddenly feared that any test they didn't order might one day come back to haunt them in court. Health care costs skyrocketed.

The nation's earliest effort to deal with the rising cost of health care led to the invention of managed care. Under the managed care system, patients were assigned to a pool of physicians practicing under contractual agreement with the patient's insurance company. Physicians were paid a fixed monthly premium per patient, regardless of how much care any one patient required. These agreements also limited the patient's care to a specific, predetermined set of protocols.

By establishing the insurer as the ultimate arbiter of care, managed care deprived both doctor and patient of a voice in choosing treatment strategies and devastated the traditional partnership between them.

The managed care concept swept the nation like a typhoon, threatening to leave adrift any patient or physician who didn't play along. Soon after I bought the clinics, a letter arrived from the state informing us that patients receiving care under existing welfare and disability programs would no longer be reimbursed for treatment unless we too adopted a managed care approach.

Unwilling to abandon our poorest patients, we decided to use the state's managed care demand as an opportunity to implement my dream of a truly integrated approach to health care.

We were encouraged by the public's growing interest in alternative therapies like acupuncture, biofeedback, chiropractic care and herbal medicine. While the conventional medical establishment quickly dismissed anything unfamiliar as quackery, our patients personally experienced the benefits of these therapies and saw things differently. Complementary and alternative medicine not only worked, it was often more cost-effective than conventional treatment strategies.

Since few physicians offered these services and most health insurance companies refused to pay for them, these options were not always available to patients who needed them. In 1995, the state insurance commissioner tried to address the problem by promoting legislation requiring insurance companies operating within the state of Washington to provide coverage for alternative therapies. The law passed and went into effect the following year. We were way ahead of her.

We had already assembled a team of family practitioners, pediatricians and gynecologists working side by side with naturopaths, massage therapists, biofeedback therapists, physical therapists, acupuncturists, chiropractors and nutrition therapists under a single roof. This diverse staff roster allowed our patients to choose the health care strategy that was best for them.

At our peak, the clinics operated at six primary locations, with six additional satellite branches in Seattle and throughout the Puget Sound region. To better serve the public, our clinics remained open seven days a week and required no scheduling of appointments.

Better still, our new system was affordable. Complementary care and a well-orchestrated team approach allowed us to offer clients a comprehensive, full-service managed care plan covering everything from major in-hospital surgical procedures to simple in-clinic wellness care for just $115 per month—roughly one-third the cost of policies offered by conventional insurance companies.

The clinics and managed care pool both grew quickly. Whether patients chose to remain with their employer-provided health insurance or took advantage of our managed care plan, they all loved our new system. Our innovative approach to care worked so well that I began getting calls from hospital administrators asking why our patients weren't showing up in emergency rooms and seldom required surgery.

The answer was obvious—at least to me. A team of practitioners from diverse backgrounds and traditions pulled together to offer a unique system of care designed to meet the needs of the individual. Our customized, wellness-oriented approach to medicine meant healthier, happier patients. Healthier patients meant lower costs.

Before long, our clinics provided services for ninety thousand patient visits each year. The astonishing success of our low-cost integrative care strategy convinced us that it was time to extend our care network nationwide with an initial offering of public stock. This made everyone happy—except the insurance companies, with whom we were now in direct competition.

By the early 1990s, I had begun to notice a disturbing trend not only among my patients, but among the public at large. Americans were getting fatter—a lot fatter. Because obesity was invariably accompanied by a whole host of deadly maladies like diabetes and high blood pressure, I found this deeply troubling and began working aggressively to do something about it.

It was a proactive strategy: if we could spot problems early, we could step in with simple, effective solutions before our patients were in real trouble. We introduced nutrition therapy and exercise to the care regimens of our overweight patients and monitored weight as a potential risk factor. We tried to make a difference in their lives before they wound up on a gurney—or on a slab.

The decision to include weight management in our medical services made me a pioneer in my field, which is not always a safe place to be. Because few of my colleagues were taking the obesity epidemic seriously in those days, the insurance industry dismissed it as a cosmetic problem unworthy of medical treatment or attention.

Today the entire medical community recognizes obesity as a killer, even among our children. Obesity is consistently associated with diabetes, high blood pressure, high cholesterol and high mortality. It is now universally recognized as one of the leading medical risk factors of our time.

Twenty years ago—when I first began addressing the problem—things were different. In short, our philosophy of care was so far ahead of its time that we found it nearly impossible to explain to those accustomed to doing things the old way. Noting my use of the word "obese" on the charts of patients suffering from diabetes and high blood pressure, in 1997 insurance industry watchdogs leveled

accusations of fraud and instigated a legal battle that would drag on for years.

We spent countless hours and unimaginable amounts of money trying to explain that treating obesity early would save lives and reduce the overall cost of health care. The insurance industry didn't see it that way; we were accused of running "fat farms."

We spent years trying to explain our team approach to care too. A patient suffering from severe back pain might be treated by a massage therapist, a biofeedback therapist and an acupuncturist working together under the supervision of an orthopedist. While recently passed state law required insurers to pay for these services, the health care industry had not yet created a standardized method of billing for them, so each provider was left to design his or her own. Given our team approach to care, it seemed logical to have our team leaders—qualified M.D.s—sign off on the bottom line of treatment authorizations and billing forms. Ironically, insurance companies whose employees initially advised us to use this approach later refused to recognize the claims and accused us of billing fraud.

The legal wrangling over complex medical issues continued for years before ever going to court. In early 1997, the insurance companies worked with the United States Department of Justice, raising the possibility of federal charges under newly passed HIPPA regulations. However preposterous this seemed, it still left us facing an unenlightened legal opponent with endless time, unlimited resources and bottomless pockets.

Determined to prove my innocence, I even took and passed a lie detector test. But innocent or not, legal fees continued to mount. Our six clinics closed one by one, leaving one hundred fifty doctors and staff members unemployed and thirty thousand active patients without access to their doctors. We could do little but watch in numb disbelief while lawyers wrangled behind closed doors and everything we worked twenty long years to build crumbled around us. When everything was gone, we had no choice but to declare personal and corporate bankruptcy.

It was a living nightmare. After four exhausting years, my eminently sensible wife finally pointed out that even prison would be better than what we were going through and urged me to accept a plea bargain agreement with the U.S. Department of Justice.

She was right. Four years of unremitting stress took a horrific toll on my health. I also had to admit that no matter how much I wanted to continue the fight, we had nothing left to fight with. If we fought on without resources and lost in court, I could be facing as many as ten years in prison, and our two small children would grow up without a father.

## CLOSING THE CIRCLE

Life holds magic in the most unexpected places; I found much of it in saying my final farewells to family, friends and adversaries.

Humans have a powerful emotional incentive to trust experts, especially when we are afraid. We want to believe that they are wise and right and all-knowing—because if they are, then they might also be clever enough

*My Prison Sanctuary*

medical wisdom and indulged my instinctive craving for the warm sands and radiant sunlight of Mazatlán, I probably would have died — and proved that cardiologist right.

Remember the children's party game Telephone? A group of children sit quietly in a circle while an adult whispers a secret message to the first child. The secret is then passed from child to child, always in a whisper, all the way around the circle until the final player says it aloud for all to hear. The game usually ends in a fit of giggles because the message has changed so much in transmission.

to save us from the things that frighten us. This desire for reassurance is so powerful that we will ignore our own instincts to obey it. We seldom stop to ask if the opinions of our experts are actually correct. This is a mistake.

Patients often cling tenaciously to the first diagnosis uttered by someone in a lab coat, too frightened to ask whether that opinion is accurate or not. When a cardiologist told me I was dying, I assumed he was right and went home to die. If I hadn't ignored conventional

Our judicial system precludes face-to-face interactions and resolutions. No direct contact takes place between the parties involved, and all information passing between them is filtered, framed and phrased by many layers of intermediaries. Because of this our legal and

*Wellness at Warp Speed*

medical encounters are inefficient and costly.

When faced with a complex legal challenge I did not understand, my initial instinct was to arrange a meeting with my adversaries. I somehow just *knew* that talking face-to-face would allow us to resolve our dispute simply and amicably. In retrospect, I wish I had listened. I deferred instead to the expertise of my attorneys, who repeatedly assured me that such a meeting would be a very bad idea.

It was bad advice. When it finally became clear that years of attempting to communicate through two opposing teams of lawyers had failed utterly and completely, I insisted that my attorneys arrange the meeting my intuition had called for from the beginning. Before consenting to a plea bargain agreement, I wanted to meet the members of the federal prosecution team. I wanted them to meet the man they were prosecuting.

I used the last-minute gathering as an opportunity to tell my story, unvarnished and unfiltered. When I'd finished, the head of the criminal division of the U.S. Department of Justice spoke up. Noting that my account had been both touching and of concern to him, he acknowledged that I might be regarded as a health care pioneer in the future, but current law compelled him to prosecute me according to existing statutes.

I couldn't help but wonder how things would have turned out if we'd talked years before. I would later come to recognize the value of this difficult experience. It helped me to understand and trust my own initial instinct, which is to face my problems directly, openly and without interference.

The following day I pled guilty to two charges of conspiracy to commit health care fraud and was sentenced to thirty-five months in a U.S. federal prison camp.

I was deeply touched by the outpouring of support from my patients. Many wrote letters to the judge pleading for leniency on my behalf.

> Prison taught me the power of love, gratitude and forgiveness—the emotional states that open our energetic pathways to the power of quantum healing.

Others came to the courthouse in person and wept openly at my sentencing.

Some came to express their condolences privately. One of these was Gary, a wheelchair-bound quadriplegic who'd been a loyal patient for more than twenty years. Gary arranged a final visit in December of 2000, just before I left for prison.

"Doc," he began, "I know your attorneys say you will be there for three years, but I want you to be ready to get out in only one."

"Thanks for your optimism, Gary," I replied in polite disbelief, "but the judge has sentenced me to three years, and from what I've been told about the feds, I'll be serving most of it."

Gary was quick with a comeback. "Think what you will, Doc, but I'm here to tell you what I see. You'll be out in one, and you should begin planning for it today."

I'd heard rumors of Gary's clairvoyant abilities but had never seen him in action. He settled back into his wheelchair.

"Are you serious?" I asked, incredulous.

"Yes, I'm serious," he replied earnestly.

I'd never seen Gary so animated before. Mystified by the odd encounter and eager to put it behind me, I thanked him for sharing his vision and hurried to write the prescription for his refills.

I thanked Gary and gave him a parting hug, not knowing when we would meet again. I didn't share his strange prediction with anyone.

## THE PRISON PARADOX

When I was in my early twenties, I had a recurring vision in which I saw myself as a forty-five-year-old physician bidding farewell to my wife, children and patients to embark on a pilgrimage to a mysterious monastery. The sense of certainty that accompanied the vision was so overwhelming that I made several journeys to India and Southeast Asia in search of the monastery that would one day be my home.

The morning of my sentencing, I suddenly understood the vision. My monastic days would be spent at a federal prison camp in Sheridan, Oregon.

I arrived at the prison an angry man, and what I found there made me angrier still. Prison opened my eyes to an aspect of reality I would never have encountered as a physician practicing in the outside world. Prison revealed the underbelly

*Suffering—whether in a hospital room or a prison cell—allows us to confront fears and overcome them, to find within ourselves the courage and heroism to become an inspiratation to ourselves and others.*

of America—a country I'd come to love and respect. Story after story echoed my own, tales of ordinary human beings swept up in the nightmare of an overpowering judicial system. In prison I met young men of almost puppy-like innocence serving long sentences for relatively minor infractions. Many of these castaways languish in prison for years, less guilty of an actual crime than of the inability to pay for adequate legal representation.

I also met a surprising number of individuals from the other end of the spectrum—prodigies whose intelligence and level of achievement dwarfed those of most living on the outside. After hearing the same disturbing tale recounted by dozens of seemingly ethical individuals, I was left with the uneasy feeling that things in America were not quite as simple as I'd once believed them to be. We pride ourselves in being a culture of hard work and achievement, but my time in prison left me wondering if there exists a magic number that triggers alarm bells in the minds of federal prosecutors—some dollar amount beyond which success is seen as a revenue source worthy of confiscation via the legal system.[5]

However diverse their backgrounds, what all these men now held in common was suffering. As a doctor, I was unable to intervene because federal law prohibits physician inmates from providing medical care or advice while incarcerated. The camp administrator made it clear that any infraction would result in a stint in solitary confinement. It was maddening to stand back and watch relatively minor health problems deteriorate into serious conditions for lack of adequate rest, nutrition and medical care.

It took the wisdom of a stranger to rouse me from my anger and awaken me to the truth. One day not long after my arrival in prison, a veteran inmate looked at me across the lunch table and said, "Doc, you're never going to

survive in here with that much anger." And then he was gone.

I never saw the man again, but his words changed my life. Awareness comes to us in surprising ways and in surprising places. It begins with the recognition of who and what we are, and with an honest assessment of the physical and emotional state in which we find ourselves. Awareness is a "You are Here" sign—the starting point on any map.

My brief encounter with the prison sage taught me that much of life's darkness comes from clouds in our own perception. Our most challenging times are not without purpose—they are seasons of growth. Impossible situations force us to stretch beyond our current limitations, sending roots and branches in new directions to find sources of light and life-giving waters once beyond our grasp.

Suffering—whether in a hospital room or a prison cell—allows us to confront fears and overcome them, to find within ourselves the courage and heroism to become an inspiration to ourselves and others.

It is good to remember that our experiences

---

[5] See Gene Healy, ed., *Go Directly to Jail: The Criminalization of Almost Everything* (Cato Institute: Washington, D.C., 2004)

make us who *we* are—but it is even more important to understand that it is we who make our experiences what *they* are. This principle is much more than just feel-good philosophy—the fact is, a century of quantum science has given us a radical new understanding of the nature of reality. The things we perceive as real—including conditions affecting our own lives—do not exist as the fixed and unchangeable solids we once imagined them to be. However strange and incomprehensible it might seem, we now know that everything exists in a vague, amorphous state of potentiality—*except when called into "solid" reality by interaction with the consciousness of an observer.*

This is, not surprisingly, uncannily similar to the description of reality handed down to us by mystics, miracle workers and healers throughout the ages. Whether by happenstance or divine inspiration, these individuals understood reality in ways that allowed them to transcend ordinary human boundaries and achieve things most people think of as impossible. That knowledge is now available to the rest of us.

Three months after my arrival at prison camp, the first of Gary's strange predictions—the one about my release date—proved eerily correct when my sentence was reduced to thirteen months. Prison was my year of miracles. By the time I left, I understood why my heroes—men like Mahatma Gandhi, Martin Luther King Jr., and Nelson Mandela—had transformational experiences in prison.

Prison strips away the chatter and demands of daily life. The solitude I found there allowed me to read, write, and pursue my research uninterrupted by an endless stream of phone calls and business meetings. Mornings began not with an overflowing inbox but with prayer, meditation and time to reflect. I'd found my monastery, and in it a lifestyle much like that embraced by monks and ascetics throughout the ages. Like them, I learned to forgive my enemies.

Prison taught me the power of love, gratitude and forgiveness—the emotional states that open our energetic pathways to the power of quantum healing. It taught me too that none of the events in our lives are without meaning. Though we may not immediately recognize the role we play in the greater web of events, our presence is vital, and it does not go unnoticed.

I received an unexpected visit from an out-of-state attorney during my stay. He'd come a long way to see me because he had something important to say: "Doc, I just wanted to let you know that you've won."

I didn't feel much like a winner sitting there in a prison uniform, but I was curious and waited for him to go on.

When he finished his story, I knew that he was right. In some small way I had won—and so had many others. Patients outraged by our situation had filed a class action suit against insurance companies operating in the state of Washington for denial of services. Several insurance companies paid in excess of $30 million in an out-of-court settlement for their failure to provide coverage for complementary and alternative medical care—the kind of care I'd been sent to prison for providing. In ways we could not have imagined, we changed the face of health care for the benefit of our patients. This costly lesson ensured that patients in the state of Washington would finally have

access to the kind of integrated health care I believed in.

It was a comforting thought.

As my perception of my circumstances changed, so too did my circumstances. I was a better man when I left prison—but I also left behind a better prison. As I was making final preparations to leave, prison staff members took me aside privately and asked me to make room for them in my future plans.

I'd be honored. There's room for all of us.

In the past few years I've traveled the world sharing a message of quantum hope and healing with audiences everywhere from Thailand to Tehran, from Denver to Dubai. In every corner of the world, the heartwarming response has been the same: We are ready to transform our lives. We are ready to transform our world. We are ready.

I hope you're ready too. If you are, let's begin our exploration of the exciting realm of love, hope and quantum healing.

It's your realm too.

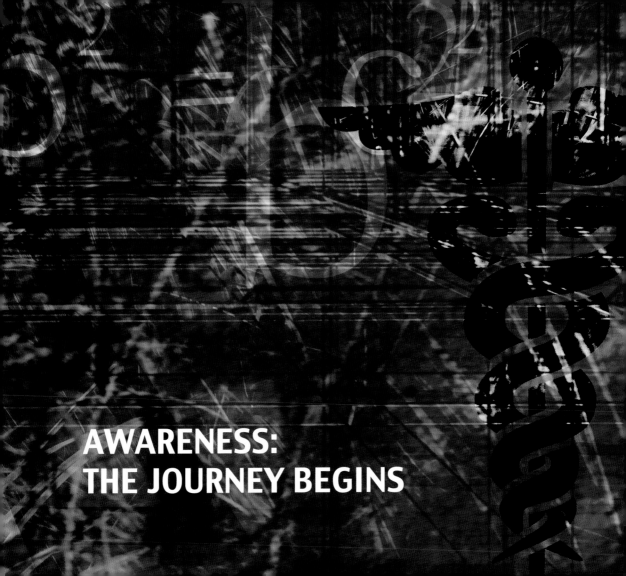

# AWARENESS:
# THE JOURNEY BEGINS

## WHERE WE COME FROM

When our earliest ancestors looked to the night sky with a sense of awe and wonder and saw in the turning of the stars patterns coinciding with the turning of the seasons, science was born. On some deep, instinctive level, these early observers must have already grasped the fundamental principle of wholeness and unity underlying all of creation. It was their innate understanding of "the way things are"—*and their willingness to act on that understanding*—that prepared our species for the cognitive leap that allowed us to begin measuring and recording our observations. These early records—however primitive—represent mankind's first formal attempts at data analysis.

Data analysis brought predictability; predictability inspired technological advances. Hunters could anticipate the seasonal migration of herds; crops could be planted on schedule to ensure maximum yield. The keen intuitive observations of these early scientists allowed our species to bridge the chasm between instinct and intellect, between observation and practical application.

Their work spawned the development of technologies that would help ensure the survival of mankind. It is not surprising that those who held the secrets of those technologies—priests, shamans and healers—would be viewed by their people with an almost mystical reverence. A little of that mystique still clings to doctors, scientists and religious leaders even today.

Knowledge, in short, had become power.

Mankind's cultural and scientific heritage grew with the coming of each generation, passing from individual to individual and clan to clan as tribes met, traded, married and warred. When nomadic tribes coalesced into fixed civilizations, libraries and centers of learning appeared on the world scene— a development made possible by another revolutionary new technology: writing.

The invention of writing allowed our ancestors to reliably transmit information across time and distance, but this advantage was not without cost. Literacy—the ability to derive meaning from an agreed-upon set of abstract symbols—further enhanced the mystique of educated "insiders" who possessed the skill. It also granted them exclusive access to mankind's growing treasury of knowledge.

An emerging class of experts was growing in prestige. The rest of us were being slowly conditioned to distrust our own innate wisdom and power of observation.

The passage of time would reveal another insidious flaw in the system: when humanity's store of knowledge is passed from one generation to the next, the process of direct personal observation is supplanted by one of indoctrination into a set of widely held beliefs… *which may or may not be true.*

When the quest for truth is sacrificed to orthodoxy, we all lose. We make progress only when those of us who question and observe for ourselves discover errors in prevailing conventions and step forward to challenge them.

> The keen intuitive observations of early scientists allowed our species to bridge the chasm between instinct and intellect, between observation and practical application.

Having spent a year in federal prison myself, I understand very well the ridicule and persecution faced by pioneers who challenge the prevailing scientific belief system—but I know too that truth has a power and a destiny of its own. Once spoken aloud, it shines, quietly attracting supporters until eventually—perhaps a year, a decade or a generation later—an open split occurs within the scientific community. The battles that follow are often brutal and protracted, but once open debate and serious investigation begin, the outcome is inevitable. In the end, the truth will prevail.

## PIONEERS

Long ago, in the distant mists of time, the skills of a Greek healer and physician by the name of Asclepius were so highly regarded that he reached mythological status as the god of medicine. While we know little

of Asclepius the man, his profound impact on the history of medicine is beyond question. Hundreds of Asclepian healing temples dotted the ancient world—the legacy of the great healer and his heirs—among whom were Hygieia and Panacea.

Physicians practicing in Asclepian temples adopted his whole-being approach to wellness, working to optimize the physical, mental, emotional and spiritual health of their patients. Sick and injured pilgrims traveled for miles to sleep on the temple's sacred grounds in hopes that Asclepius would inspire a dream that would reveal their unique path to healing.

It is fitting that the staff of Asclepius came to us as the familiar serpent-entwined rod symbolizing modern medicine. Emerging evidence validates the effectiveness of integrated

*Asclepius – god of medicine*

healing and calls for a revival of the Asclepian healing tradition. Whether our generation witnesses these changes or not, the great healer's impact on the development of medicine through the work of his students has been immeasurable.

Hippocrates, known today as the Father of Medicine, was introduced to his trade at an Asclepian healing temple on the Island of Kos more than two thousand years ago in ancient Greece. Hippocrates developed his own theories of medicine and sought the cause of illness in purely physical malfunctions of the human organism.

Galen—the legendary second-century Roman physician and surgeon—also trained as an Asklepian practitioner. When Galen was a child, Asclepius visited his father in a dream to bring word of Galen's destiny as a physician. The vision proved true: Galen became one of the

The Hippocratic Oath, drafted in 400 BC, is still used by physicians today as the code of professional conduct.

Hippocrates (460 – 377 BC), the Father of Medicine and a student of the Asclepian school of healing, adopted a whole-body approach to optimize the physical, mental, emotional and spiritual well-being of his patients.

most notable physicians of all time. His reign as the leading authority on medical theory lasted fourteen hundred years.

As time passed, the wisdom of the ancient world drifted from civilization to civilization on the winds of fate and political fortune. The crumbling of the legendary Roman Empire's centralized bureaucracy set the West's intellectual heritage adrift on the four winds, which ushered in the Dark Ages. Some of it simply vanished; some of it was preserved through the works of enlightened

Muslim scholars and physicians like Abu Ali Sina (known in the West as Avicenna); some remained hidden away in churches and monasteries strewn across Europe by Christian missionaries driven to the farthest corners of the world by brutal persecution and evangelical zeal.

Throughout Europe, monks hidden away in dimly lit sanctuaries stood as the sole arbiters and interpreters of truth. Custodianship of the West's intellectual heritage granted the church great power and authority, which it guarded most jealously. Few dared to question the dictates of religious authorities; those who did often paid a heavy price.

Our species learns in waves, each crest of observation and discovery followed by a trough of indoctrination and repression. When authority is no longer questioned, science stagnates. This time, centuries passed before the process of observation emerged from the trough of indoctrination.

The spark that reignited the West's intellectual curiosity was generated by the sixteenth century Polish physician and scholar

Nicolaus Copernicus. Copernicus questioned the works of long-unchallenged authorities and realized that the Earth moves around the sun—a belief held in sharp contrast to the church's official position. Copernicus died in 1543 and so would not live to see the controversy generated by his work—nor could he prove it with direct observation. The telescope would not be invented until 1608.

The defense of Copernicus's theory fell to a seventeenth century scientist and physician, Galileo Galilei—an inveterate tinkerer with a passion for mathematics and a deep fascination with physics. When Galileo improved primitive telescope technologies enough to allow for direct observation of the planets, what he saw in the night sky convinced him that Copernicus was right.

Galileo couldn't have imagined the firestorm of historic events his observations would set into motion. He was dogged by years of controversy and eventually stood trial before the Inquisition on charges of heresy in 1633. The trial did not go well: Galileo's work was banned by the church as heretical, his books were burned, and he himself was condemned to spend the rest of his life under house arrest.

In 1992—three centuries after Galileo's trial and condemnation—Pope John Paul II finally stepped forward to express the church's regret for the mishandling of the affair.

*Canon of Medicine*

*Abu Ali Sina (980 – 1037)*

The brilliant Iranian physician, philosopher and scientist Abu Ali Sina wrote the five-volume, one-million-word *Canon of Medicine*, which remained the world's most respected textbook of medicine until the late 1700s.

Copernicus and Galileo breathed new life into mankind's quest for knowledge. Their rejection of blind authority and passion for observation inspired a revolution of greater worth than any single theory or invention. As increasing numbers of scientists realized that authorities were not infallible, mankind entered a new era of discovery.

Everything was up for grabs—including intellectual hegemony in the West. Scientists frustrated with scholasticism's blind adherence to tradition began challenging underlying assumptions about everything—and promptly put on new blinders of their own.

If Galileo was the father of modern science, the French scientist and philosopher René Descartes was the prodigal son who placed science on the pedestal it enjoys today. Descartes championed

reason as the path to understanding—a school of thought known as rationalism. A competing scientific philosophy known as empiricism held that scientific knowledge came through experience, measurement and experimentation.

Descartes introduced another sort of dualism as well: his emphasis on the distinction between mind and brain struck the blow that severed the human being into spiritual and physical components—a wound from which the field of medicine has not yet recovered.

In some respects, these debates were like arguing over whether a man would be better off with a left leg or a right, but once distinctions had been drawn, the "sides" inevitably battled for supremacy.

With the church at odds with science over the question of authority and scientists philosophically at war

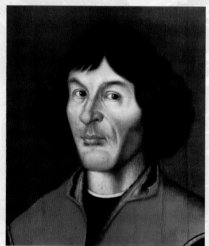

*Nicolaus Copernicus (1473 – 1543)*

among themselves over the proper scope, procedures and definition of their trade, it was a battle for the ages.

When the dust finally settled, materialism emerged victorious. Matter mattered. For many scientists it was all that mattered. Anything not measurable by existing levels of technology was viewed with skepticism; evidence that lent credence to the existence of a nonmaterial reality was dismissed as anecdotal and irrelevant, or worse, as superstitious nonsense. Scientists bold enough to investigate anomalous evidence were ridiculed by their peers and treated as a new class of untouchables.

*Galileo Galilei (1564 – 1642 )*

With entire fields of inquiry out of bounds, orthodoxy once again reigned supreme. It would remain so until the early years of the twentieth century, when pioneers working in the field of quantum physics peered through the keyhole of the tiny subatomic world and discovered a new reality vast enough to encompass both matter and spirit.

Throughout history, those with the courage to question have reminded us that the experts

*René Descartes (1596 – 1650)*

have spent the most time studying the map. Paradoxically, those who spend the most time memorizing the map often have little time left to explore the territory itself.

That task belongs to the rest of us.

Fortunately, the greatest legacy left to us by our ancient star-gazing ancestors lies not in our degree of literary advancement or technological prowess but in ourselves. We have inherited from them the power of observation, the ability to learn from our experiences, and a deep, innate understanding of "the way things are." The question we must each confront in life—especially in times of illness—is this: *Am I willing to act on that understanding?*

Doctors are experts, but *you* are the world's leading authority on your own body. And when it comes to your health and happiness, you have the most at stake. Don't be afraid to challenge what you've been told by others. Explore new alternatives, research all of your options and choose the ones that are right for you. Expect

are sometimes wrong.

The fact is, we in the modern world have inherited a vast body of knowledge from our predecessors, and that body of knowledge represents humanity's best attempt to date at drawing a map of reality.

"Experts" in any field—including medicine—are by definition those who

> Cogito ergo sum.
> I think, therefore I am.
>
> – René Descartes

respectful answers to your questions. If you don't get them, find a new doctor.

It's not impolite. It's just smart.

## WHY IT MATTERS

Science, health care and societies all share a common pattern of learning and discovery. New discoveries are comparatively rare because most research efforts are devoted to gaining a better understanding of existing concepts. While efforts to better understand the details of an existing theory can yield useful information, our greatest gains come from leaps of intuitive insight that occur when unconventional thinkers set aside current assumptions and reconsider old problems in a new light.

*Dr. Albert Schweitzer*
*(1875 – 1965)*

Conversely, when professionals—especially those working in the field of medicine—stop questioning their underlying assumptions, things go horribly wrong. When things do go wrong, they stay that way for a very long time because few outside the profession are willing to question the presumptions of an entire body of "experts."

The history of medicine has been tragically marred by a stubborn lack of awareness and innovation on the part of medical professionals. The costs have been steep, and they have inevitably been paid by ordinary people who deserved better.

In the nineteenth century, no one believed that tiny, infectious microorganisms could be the cause of fevers, chills and death—not even doctors. Most physicians were so entrenched in existing medical theory that they had lost the capacity to think and act independently. Dr. Ignaz Semmelweis was a notable exception.

Heart disease and cancer are currently the leading causes of death among Western women—but that has not always been the case. That unenviable title was once held by a disease called puerperal fever, a dreaded ailment that killed one woman in six, usually within weeks of giving birth. Nearly a third of the women giving birth in medical facilities died of "childbed fever"; women giving birth at home remained largely unaffected.

In 1846, the prestigious Vienna General Hospital housed two separate obstetrical clinics. Women giving birth at the first clinic—a full-service teaching facility for medical students—had a childbed fever mortality rate of roughly 9.9%. Women giving birth at the second clinic—a teaching facility for midwives—fared much better. At the second clinic the childbed fever mortality rate was just 3.4%.

When Dr. Semmelweis took a position as assistant physician of the dreaded First Obstetrical Division, he took it upon himself to begin investigating the cause of childbed fever. His superiors—who believed the disease to be

Dr. Ignaz Philipp Semmelweis (1818 – 1865)

nonpreventable—objected to his efforts. They objected still more when Semmelweis put two and two together and concluded that medical students performing autopsies in the cadaver lab were carrying a contaminating agent to the obstetrical ward on their hands.

When Dr. Semmelweis implemented a department-wide policy of handwashing between autopsy work and patient exams, the overall mortality rate due to childbed fever dropped to just over 1%.

Despite Dr. Semmelweis's success in virtually eliminating childbed fever from the wards, his ideas were not well received by his colleagues. Since the medical profession did not even acknowledge the existence of germs, it could not justify implementing procedures to control them—no matter how effective those procedures might be. Doing so would be a concession to sloppy science.

Three decades later, experiments conducted by Louis Pasteur conclusively proved the existence of germs. Pasteur's work inspired Joseph Lister's ongoing research and contributed to the discovery of antiseptics. In 1879, Pasteur identified a specific strain of bacteria in the blood of women with childbed fever.

In spite of the evidence provided by these pioneering physicians, their work did not lead to widespread changes in medical practice until the beginning of the twentieth century.

Meanwhile, women waited—and died by the thousands.

Dr. Semmelweis would not live to see the changes his work inspired. He died, impoverished and unrecognized, on August 13, 1865 in an insane asylum in Vienna.

If you ever find yourself wandering through an old cemetery, take a closer look at the tombstones. Chances are you'll find a disproportionate number of young women buried there. Some of them died of childbed fever.

The story's all too poignant moral is this: women who died of puerperal fever between Dr. Semmelweis's epiphany in 1848 and the widespread adoption of antiseptics at the turn of the twentieth century didn't really die of a disease—they died of a blind spot.

It would be easy for us to condemn my predecessors for their blind, stubborn lack of awareness—but doing so would distract us from a more important point:

*Humanity's understanding in any field of endeavor is always in a process of completion, never complete. No matter how advanced our level of knowledge, we will always have more to learn. Despite the tremendous advances made in the last century by those in the medical profession, blind spots remain—even today.*

*The most dangerous blind spots of all are the ones we don't know we have.*

> It's supposed to be a secret, but I'll tell you anyway. We doctors do nothing. We only help and encourage the doctor within.
>
> – Dr. Albert Schweitzer

## "YOU ARE HERE"

Despite modern surgical and pharmaceutical intervention, mortality tables tell us that the vast majority of people will die of heart disease or cancer. Farther down the list of horrors we find vascular diseases, respiratory diseases and a host of other unpleasant maladies that stem from an imbalance in critical physiologic systems or an immune system operating at less than peak capacity. Violence—whether inflicted by ourselves or others—is there too.

One of the surest routes to being branded a quack is to suggest that disparate medical conditions share a single causative agent or contributing factor. Nonetheless, I am going to do just that, because I am convinced that we are facing a situation similar to that which killed thousands of healthy young women in the latter part of the nineteenth century.

Early physicians simply could not (would not?) believe that their lack of hygiene aided and abetted tiny, infectious hitchhikers so small as to be invisible yet powerful enough to kill a full-size human being. No matter what the facts said, doctors refused to wash their hands and instruments for nearly half a century because the truth was simply too preposterous to believe.

This time it's our turn. Once again, we are facing a threat that is invisible, transmissible, lethal—and almost entirely unrecognized by the medical community.

Decades of research have linked heart disease, cancer, diabetes, Alzheimer's, obesity, ulcers, depression, insomnia, drug and alcohol abuse, suicide and every other disease known to man to our mental and emotional states.

Some of these conditions wreak such immediate and overwhelming havoc on the body that death occurs at once. Others are more subtle, inflicting cumulative damage over a period of time until the whole system collapses. Either way, the results are the same.

There is nothing new about any of this. Doctors have known about the mind-body link for decades, but this knowledge had little impact on the lives and well-being of patients.

Centuries of medical tradition have conditioned health care practitioners to view the human body in purely mechanical terms. Each organ is viewed as a discrete mechanical entity. Each disease is viewed as a purely physical malfunction of that mechanical entity. The hyperspecialization that resulted from this kind of thinking blinded many in my profession to the interrelated nature and function of every cell in the human body.

If today's specialists are failing to address even the obvious physical interconnectedness of the human body, it should not surprise us that the link between thought, emotion and physical wellbeing is never dealt with at all. Like Dr. Semmelweiss' colleagues, many of today's physicians witness the effects of emotion on health on a daily basis but remain blind to their cause. We are so thoroughly conditioned to think in physical terms that we cannot imagine an agent powerful enough to damage the patient yet invisible to current lab technologies.

In thirty years things will be different. Much as the pioneering research of Lister and Pasteur paved the way for lifesaving changes in medical practice, our generation too has its pioneers—but this time the research that will save the lives of countless patients is being done in physics labs.

Recent discoveries in the field of quantum physics have introduced us to levels of interconnectedness that extend far beyond physical systems, beyond even the mind-body-spirit link today's practitioners find so difficult to grasp. In the future, physicians will use these levels of interconnectedness routinely as everyday tools of the trade.

But what about the rest of us? What are we to do if we would rather not wait decades for medical practice to catch up with theory? What if we've been told we may not have thirty years to wait?

## THE WAY FORWARD:
## LESSONS OF THE HEART

**G**ood news! You don't have to wait. You don't have to wait at all. Your journey to wellness can begin right here and now.

In February of 2005, researchers at Johns Hopkins published the results of a study on a recently identified condition called stress cardiomyopathy. The condition mirrors the effects of a heart attack—chest pain, shortness of breath, fluid in the lungs and heart failure—and can be every bit as fatal.

The difference? A classic heart attack results from impaired blood flow to the heart muscle. If blood flow is not restored, the oxygen-starved portion of the heart muscle will deteriorate and eventually die. If a sufficiently large section of the heart muscle dies, so will the patient.

While heart attacks can be triggered by emotional shocks, stress cardiomyopathy—also known as "broken heart syndrome"—is caused exclusively by sudden and intense emotional stress and can occur even in healthy individuals with few risk factors and no evidence of arterial blockage. The syndrome occurs when an intense emotional shock sends a massive surge of catechol-amines (stress hormones like adrenalin and noradrenalin) pouring into the bloodstream, creating a self-brewed "toxic soup" powerful enough to stun the heart.

Physicians have struggled from medicine's earliest days to understand the heart and define its role and function. Aristotle, the preeminent biologist of the ancient world, saw the heart as the very center of human physiology, a furnace in which burned the sacred flame of life. Later, in the age of rationalistic materialism, the brain—the center of reason—ascended the throne, and the heart was demoted to a simple mechanical pump.

We now know that in a sense they were both right—and they both missed the mark entirely. Reductionist thinking does not always yield the truth.

The heart *is* a pump. The heart is also a complex sensory organ. The heart is a communications device—a modem, if you will—that continuously exchanges data with the world around us and with the greater quantum realm beyond.

The human body is not a collection of competing parts—it is a fully integrated system operating within and inseparable from the greater quantum field.

Your heart, body and mind work together as *a fully integrated system* to process information and formulate a response to the things happening within and around you. Because of this, your thoughts, feelings, beliefs and attitudes can kill you—or they can make you well. The recovery process is not random or beyond your control; in a very real sense, the choice is yours.

If you're ready to take charge of your life, stay tuned. You are about to discover powerful techniques, information and strategies that will help you create the future of your dreams.

QUANTUM WHAT?

## INDIVISIBLE UNIVERSE

Epiphanies happen every day, but most of them slip by unnoticed because we are preoccuppied with the demands and distractions of everyday life. Illness can change all of that, and it can do it in an instant.

If you want to transform your health, begin by transforming your thoughts. Convalescence can be a wonderful opportunity to start that process. In 1989 I was lying in bed feebly breathing oxygen through a nasal cannula when the author Michael Talbot revolutionized my way of thinking. In the book *The Holographic Universe*[4], Talbot graphically and emphatically describes the nature of our universe as a seamless, indivisible whole. His words set me on the path to recovery from an incurable and life-threatening illness.

Talbot's vivid literary representation of David Bohm's "indivisible universe" and John Bell's Theorem of Non-locality hit me like a ton of bricks. It was the first scientific evidence I found hinting that my diagnosis of viral cardiomyopathy might not be a death sentence. That I somehow had direct control over the events taking shape in my own heart seemed as

*David Bohm (1917 – 1992)*

[4]Talbot, Michael, *The Holographic Universe* (New York, HarperCollins Publishers Inc., 1991)

strange to me as the idea of Martians invading New York City.

Existing in a continuum implies a connection to everything. Mind is not separate from heart. They are part of the same continuum. Therefore, I could exert control over my heart by changing the direction of my thoughts. I was quite familiar with similar research in the emerging field of mind-body medicine, but this was something new. This was the science explaining the how and why of mind-body medicine's success.

Talbot had beautifully articulated the work of two geniuses who'd unlocked the mystery of the human body and revealed a secret passage toward the Fountain of Youth. This was the closest thing to the holy grail of medicine that I could imagine. David Bohm and John Stewart Bell had boldly gone where no one in modern medicine had dared to venture. They extended Albert Einstein's idea of the space-time continuum to include everything in the entire universe. It was all part of a single continuum—and that powerful continuum included everything I would ever need to bring about a complete cure in my own body.

Despite the fact that I am not a physicist, the epiphany that occurred the instant I grasped these quantum principles empowered me to use the theories of Einstein, Bohm and Bell to restore my body to health.

It can do the same for you.

## NON-LOCALITY

My introduction to non-locality came in 1989 as I lay tethered to an oxygen tank and fighting for every breath. It was difficult to keep hope alive in the face of my bleak medical prognosis and rapidly deteriorating physical condition.

> Mind and matter are not separate substances. Rather, they are different aspects of one whole and unbroken movement.
>
> – David Bohm

It was exhausting to think of my sudden physical collapse and the limited options available to me in the remaining months of my life. I understood and empathized with the millions of people diagnosed with terminal illnesses who lose hope and wish for a quick and easy way out.

If you are one of them, don't quit just yet—your future may be brighter than you think. Non-locality changed everything in my life. It can change yours, too.

As a physician, I had read countless accounts of patients who'd used mind-body exercises to trigger spontaneous healing in their bodies. Like most of my colleagues, I'd dismissed these cures as anomalous or anecdotal because I'd never seen convincing scientific research explaining the mechanism at work behind them. I would later learn that the evidence had been there all along, but my years of training and conditioning as a clinician had become an obstacle to understanding non-locality and its role in everyday healing.

Knowing that my own death was imminent inspired me to take a closer look. Would it really be possible to increase the flow of oxygen to my heart muscles and the rest of my body by focusing exclusively on thoughts of healing? I wanted to believe that I could cheat death and heal my heart with my thoughts alone—but I also wanted rational scientific proof that such a thing was possible. I needed a physician—better yet, a physician-scientist—who could show me how to cure my ailing heart with a scientifically proven mind-body technique.

Mathematical proof for the existence of non-local interactions among subatomic particles surfaced in 1964 when John Stewart Bell showed how two different particles separated by distance and time remain unified and inter-connected. The technology available to Bell at the time wasn't sufficiently advanced to allow him to prove his hypothesis in a laboratory. In 1982, physicists Alain Aspect, Jean Dalibard and Gerard Roger of the University of Paris validated Bell's work experimentally.

The French team produced a series of twin photons by heating calcium atoms with lasers. These photons (particles of light) were then sent traveling in opposite directions through 6.5 meters of pipe. In the course of their journey, the photons passed through filters that directed them toward two possible polarization analyzers. Each filter took ten billionths of a second to switch from one analyzer to the next—precisely *thirty billionths of a second less than it would take for light to travel the distance that now lay between the particles.* The instantaneous nature of this connection ruled out any possibility of the two photons communicating with each other through any known physical process, including the energy of sound or light itself.

Aspect's work validated Bell's mathematical prediction that each photon would correlate its angle of polarization with its twin despite being separated by distance. The effect was stunning: the two particles performed an identical "photon dance" simultaneously, instantaneously and faster than the speed of light! The two photons were separated by thirteen meters, but they were connected non-locally.

The effects of non-locality are not limited to the world of subatomic particles. This key quantum principle plays a fundamental role in all interactions on our planet and in our everyday lives. As English physicist Paul Davis of the University of Newcastle explains, "The non-local aspects of quantum systems is a general property of nature, because *all* particles are continually interacting and separating."

While you may not recognize *non-locality* by its scientific name, the principle has played an important part of life on Earth from the beginning. Non-locality explains why you sometimes reach for the phone to call a high school friend living a continent away just as she prepares to call you. It explains why your prayers can non-locally heal a friend lying in a hospital bed a thousand miles from your home. It helps us understand the morbid sense of loss and anguish felt by millions around the planet when a devastating tsunami struck beach communities on the shores of the Indian Ocean in late 2004. It explains the strong bond and sense of connectivity that exist between all life forms.

Albert Einstein (1879 – 1955)

It explains why our species still exists today, how our kind has managed to persevere despite centuries of violent turmoil and conflict. Love is the world's most powerful healing force, and now, for the first time in recorded history, we have mathematical proof of its existence and an explanation of how it works. Non-locality is the physicist's description of how love and intimacy actually operate. Like electrons, protons and neutrons, love is invisible, ubiquitous and non-local, but it has all the power we need to heal our hearts, our lives and our planet.

In 1989, I used David Bohm's and John Bell's work in non-locality to heal my physical body. Today I use my own understanding of non-locality to facilitate a deeper connection with the invisible emotional and spiritual realms. No matter where I find myself, whenever I spend time with my family, friends or community, I am constantly reminded of the ever-present and non-local healing power of love. I hope you are as well.

## THE MATTER - SPIRIT CONTINUUM

In 1905, Albert Einstein's famous $E=mc^2$ formula proved that energy and matter are not only equivalent, they are one and the same. Einstein's Theory of Relativity beautifully illustrates how energy manifests into matter and matter into energy. We live in a physical world of solid objects—cats, cars and computers—but we also inhabit the world of invisible energies that contains these physical apparitions. Every bit of matter has an energetic equivalent and every unit of energy has physical attributes.

Right now it may be impossible to imagine how you can generate these transformations, but given the right tools and appropriate conditions, you will one day be able to move from one state to the other with relative ease.

American scientists used this principle in 1945 to crack the atom and unleash the deadly force of the first nuclear bombs over Nagasaki and Hiroshima. By the end of this century, we will have learned to harness the power of the atom for purely humanitarian purposes.

New technologies will permit us to reverse illness and disability and allow us to extend the human life span by changing the energetic substrate of living tissue. I have witnessed the awesome clinical potential of a number of these pioneering technologies over the past several years. I believe these tools will soon replace our heavy reliance on surgery and pharmaceuticals.

But the technological marvels of our time will not be the only way to harness the energetic values of physical forms. Accessing the matter-spirit continuum is nothing new for the yogis of India, the Sufi dervishes of Iran or the shamans of Central America. They've used it routinely to heal and perform miracles for thousands of years.

The real question is: Can you and I access the same hidden powers and apply them in our own lives? Can we harness the quantum potential of the physical body to manifest our dreams, cure our bodies and heal others?

Twenty years ago, as a new physician just beginning the practice of medicine, I would have said no. After two decades of clinical observation and the assessment of more than 250,000 physical and mental examinations, my answer now is a resounding *yes!*

I feel embarrassed for not having grasped the significance of Einstein's Theory of Relativity earlier in my career. I received my degree in medicine from the Albert Einstein College of Medicine. I should have known better.

Einstein's formula offers mathematical proof that every chemical and biological condition, no matter how life-threatening, is reversible. In Einstein's formula $E=mc^2$, the "E" represents energy, "m" represents matter, and "$c^2$" represents the speed of light multiplied by itself.

If you write a new formula and replace "E" with "wellness" and "m" with illness, you will understand why wellness and illness represent interchangeable states of being.

Wellness and illness form a continuum and therefore exist as interchangeable states. You can move from illness to wellness as easily as you can from wellness to illness.

The mathematical blueprint for your recovery already exists. It is your destiny to find it.

# THE SCIENCE OF ENLIGHTENMENT

There's something magical about the nature of light. We're drawn like moths to the familiar neon glow of an "open" sign flickering in the window of a late-night café. Candlelight sets a mood of sanctity in our chapels—or one of romance for dinner with a special someone. We flock to beaches by the millions each year to bask in the healing rays of the sun. Visions of a starry sky still hold an almost mystical fascination for our kind, whether they come to us during a night of solitary contemplation in the desert or via televised deep-space images courtesy of the Hubble Space Telescope.

Light holds a special appeal for scientists too, but being analytical by nature, they spend a great deal of time trying to figure out what things are and how they work. In wake of the Galileo affair, scientists saw themselves as the sole remaining guardians of truth. (Who could blame them?) They took this responsibility seriously and worked hard to develop nice, orderly theories that would explain everything around them and make the world a tidy and predictable place to be.

By the end of the nineteenth century, they nearly got it—or so it seemed. In 1687, Isaac Newton summarized the behavior of the material world with three simple laws of motion. Two centuries later, James Clerk Maxwell provided missing pieces of the puzzle with four equations explaining the behavior of electromagnetic radiation. To classical physicists the world seemed a straightforward place where simple forces acted on objects in predictable ways to produce reassuringly certain results.

Then things began to happen. Unexpected things.

## DUALITY

Isaac Newton argued that light was by nature particulate; his brilliant Dutch contemporary Christian Huygens was convinced that light traveled in waves. At the turn of the nineteenth century, an English physicist by the name of Thomas Young provided an evidentiary means of settling the debate with a simple experiment.

If you ever tossed two pebbles in a pond and watched the behavior of the intersecting ripples of water, you witnessed what scientists call wave interference patterns. When two wave crests align, they blend to form a single, taller crest. When two wave troughs align, they join to form a single, deeper trough. If a crest and trough align, they cancel each other out, and the surface of the water will be flat.

Young conducted a similar demonstration with light. You can envision his experiment by imagining three screens stacked sequentially, front to back, like index cards in a file box — with space between them forming two small chambers. Young cut a narrow slit in the center of the first screen through which he could shine a beam of light into the first chamber.

He cut two narrow slits in the second screen, each equidistant from the center point. Some of the light passing through the single slit into the first chamber continued onward through the two slits in the second card into the second chamber. No surprises so far.

The interesting bit is what happened when the light reached the final screen. When light passing through the two slits in the second screen hit the back wall of the second chamber, it created a series of alternating light and dark bands. This is the precise interference pattern you would see if you had conducted the experiment with waves of water interfering with each other at the edge of a pond.

The crests of the light waves interfere with each other *constructively* to create bright bands of light; the troughs of the light waves interfere with each other *destructively* to create the dark bands.

Young's experiment proved that the light had passed through the slits in waves. If light traveled through the two slits in the form of particles, it would have behaved like pebbles being tossed through two holes in a fence. When you finish, you expect to find two bright "piles" of light on the back wall.

So Young's experiment settled it: light is a wave.

Not so fast. (This is where things start to get interesting.)

Wave theory and Maxwell's equations both predicted that when specially constructed metal chambers called blackbodies were heated, they would emit energy uniformly, with each electromagnetic wave, no matter how small, carrying some portion of the overall energy. This created some very sticky problems for classical physics. If the accepted theories were true, then blackbodies should contain an infinite amount of energy. They didn't, and no amount of mathematical wrangling on the part of classical physicists could explain why. This didn't stop them from trying, of course, and in 1900, a German by the name of Max Planck finally resolved what had come to be known as "the Ultraviolet Catastrophe" with an astonishing leap of intuition.

What Planck realized was that energy was moving about in tiny packets he dubbed *quanta*. A few years later, a young scientific renegade by the name of Einstein applied Planck's insights to another of classical physics' more intractable problems—the photoelectric effect—and concluded that electrons were being jarred loose from illuminated metal surfaces by little packets of light striking the surface like bullets. Einstein called these packets of light *photons*.

So Maxwell and Young conclusively established the wavelike nature of light. Planck and Einstein proved that light consists of particles. Amazingly, they were *all* correct!

The sudden realization that such contradictory statements could both be true marked the beginning of a scientific odyssey that would lead to a great deal of consternation and hand-wringing on the part of physicists. Before long, they were going for long walks at night and muttering things like "No, no—this cannot be." (Some of them are muttering still. You'll soon find out why.)

Young's delightfully simple two-slit experiment proved irresistible for later generations of scientists who set up similar assemblies— with one significant difference. Rather than shining a light into the assembly, they began beaming a series of photons (or any of a number of other tiny "particles") into the apparatus. We would expect light "particles" traveling through the slits to behave like stones being tossed through holes in a fence. The particles should land in two little "piles" on the back

# THOMAS YOUNG'S

## *Light passes on to screen through one slit*

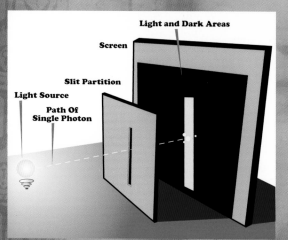

Light and Dark Areas

Screen

Slit Partition

Light Source

Path Of
Single Photon

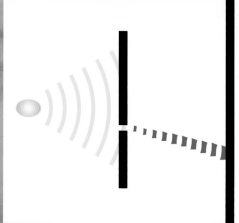

In 1801, Thomas Young designed a simple experiment to determine the nature of light.

When wave crests meet, they interfere with each other *constructively*, creating bright bands of light on the screen.

When crests meet troughs, they interfere with each other *destructively*, creating dark bands.

Light passing through the slits in Young's partition appeared in the alternating pattern of bright and dark bands characteristic of waves.

# DOUBLE SLIT EXPERIMENT

*Light passes on to screen through two slits*

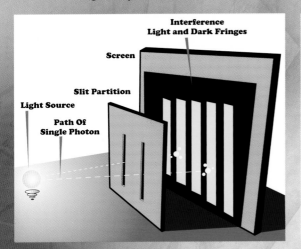

**Interference Light and Dark Fringes**

**Screen**

**Slit Partition**

**Light Source**

**Path Of Single Photon**

A century later, Einstein won a Nobel Prize for work on the photo-electric effect in which he proved light's particle-like nature. Both men were right.

This is known today as wave-particle duality.

Quantum entities like light have the potential to behave as either particles or waves.

*The observer's expectations (that means you) determine which it will be!*

wall, one directly behind each slit.

They don't.

The particles travel thorough the slits one at a time. Each lands on the back wall individually and in a separate location. But when enough of them strike the back wall, their countless tiny landings accumulate to create the distinctive interference pattern that tells us that our particles have been traveling as waves. In other words, these particles aren't passing through one slit or the other—each one is passing through both slits at the same time. Stranger still, each particle seems to *know* where the earlier particles have landed, because each one invariably lands in a spot that contributes to the emerging wave pattern.

Brace yourself. It gets weirder.

These little particles seem to know when we're watching. If we try to spy on them, they won't play. If we set up a particle detector to determine which slit they're going through, they stop behaving like waves. Each particle will pass through one hole or the other—not both. When our experiment is finished, we won't find wave interference patterns on the back wall. We'll find two little piles of particles on the back wall—one behind each slit. If we remove the detectors, the particles resume their wavelike behavior.

None of this is theory. It happens consistently, predictably, over and over again, with all sorts of particles. Quantum entities travel as waves, but they arrive as particles. Each of them seems to understand what we are trying to achieve with our experiments. Like children scurrying away from the cookie jar when Mom steps into the kitchen, quantum entities somehow know when we're watching and change their behavior when we do.

That should give you something to think about the next time you turn on a light bulb or step outside on a warm summer night to gaze at the stars. They just might be looking back.

Now you can see why the scientists who first discovered this stuff were in such a tizzy. It was their job to make sense of all this, and from the standpoint of classical physics, there was very little sense to be made. The tiny

*Werner Heisenberg (1901–1976), a German physicist and philosopher who in 1932 won the Nobel Prize in physics for creating the science of quantum mechanics.*

world of atoms and subatomic particles—*of which we are made*—defied the rules of classical physics and shattered science's long-held assumptions about the nature of our interactions with physical reality.

For starters, the double-slit experiment made it clear that the experimenter could no longer be considered an objective observer of an external reality. Light exhibits the properties of a wave. Light exhibits the properties of a particle. Which behavior it exhibits at any given moment depends on the experimenter's desires and expectations. There is no such thing as an impartial scientific observer. The experimenter changes the outcome of his experiment merely by engaging in the act of observation.

Just in case this still seems a little abstract, let's consider it in another context. Medical science has long relied on a research protocol called the double-blind study to eliminate researcher bias in experimental results. In a double-blind study, neither the participating patients nor the researchers conducting the study know which patients are receiving the new experimental drug and which are receiving a placebo. A century of quantum physics has proven that it is impossible to eliminate the effect of the experimenter on his experiment.

The experimenter in every double-blind study is subject to these quantum principles. In short, *we now know that there is no such thing as a double-blind study.*

When researchers in competing industries report conflicting study results on the risks and benefits of their respective products, we traditionally blamed sloppy lab work or economic self-interest. While we can't overlook these obvious possibilities when weighing contradictory studies, we—especially those of us working in the field of medicine—must now also consider the possibility that the experimental bias might lie at an even deeper level and may, in fact, be inseparable from the experimental process itself.

In addition to considering the impact of the observer effect on research outcomes, we need to ask ourselves an even more important question: *Who is the observer?*

If a dozen respected researchers and thousands of study participants report few side effects with a new drug entering the market, but the drug makes you violently ill, whose observations are relevant? The observations of others—including your doctor—do not play the deciding role in shaping your destiny. This is especially important to remember if you've been diagnosed with an "incurable" condition.

Your thoughts, feelings, beliefs and attitudes have a tremendous impact on your well-being because when it comes to your life, *you are the observer*. Now that you understand the power of the observer effect, you can take advantage of this dynamic quantum principle to transform your life and health.

> The violent reaction on the recent development of modern physics can only be understood when one realizes that here the foundations of physics have started moving; and that this motion has caused the feeling that the ground would be cut from science.
>
> – Werner Heisenberg

# UNCERTAINTY

In 1927, Werner Heisenberg further underscored the importance of the observer's role with his introduction of the Uncertainty Principle. In the quantum realm, paired sets of properties called conjugate variables exist that cannot be *measured* simultaneously because *they do not exist simultaneously*. It is impossible to measure both the position and momentum of a quantum entity. If you pin down an electron's location with a measurement, its velocity becomes indistinct. If you measure its velocity, its position dissolves into a haze of uncertainty.

When it comes to quantum measurement, energy and time are also mutually exclusive properties. It is possible at any given moment to determine one or the other but not both. This limitation does not stem from a lack of precision in the measuring apparatus—it's just another of the strange attributes of the quantum realm.

According to Heisenberg's Uncertainty Principle, once the observer chooses—*we choose*—which attribute to bring into focus, the rest of the picture dissolves into uncertainty. Again we see that the observer changes the thing she is observing just by deciding to look.

The Uncertainty Principle also reminds us of another important fact. As Heisenberg himself put it, "We *cannot* know, as a matter of principle, the present in all its details."

No one—no matter how impressive his credentials or level of expertise—knows everything about your condition or the outcome of a suggested treatment plan. Your health care provider can be an invaluable source of support and information, but even after your doctor has contributed everything she has to offer, *you will not have all of the pieces of the puzzle*. A substantial degree of uncertainty is hardwired into every system.

Others may provide you with enough information to assemble part of your wellness puzzle, but ultimately your own instincts, beliefs and intuition will determine the shape of your future. Once you envision your desired wellness outcome, it's time to begin looking for

the missing pieces and dis-
carding those that don't fit.

## QUANTUM SPIRIT

If quantum physics has taken from us our childlike belief in a neatly predictable world of cause and effect, it has rewarded us with the very adult pleasure of living in an exciting new realm of infinite possibility in which we have the potential to shape our own destiny. While scientists may disagree about how to interpret and frame the bizarre facts of quantum physics, one thing is crystal clear: the realm quantum physicists have been mapping for just over a century bears an uncanny resemblance to the territory described throughout the ages by the founders of the world's great religions.

The Christian New Testament attributes

*Ahmad Soleimanniyeh as Jesus in the feature film and television series "**Jesus, Spirit of God**", by Iranian filmmaker Nader Talebzadeh.*

many miraculous healings to the work of Jesus Christ, but what may be overlooked by the casual reader is the fact that Jesus himself declined credit for these healing events. While the casual reader may be tempted to dismiss Jesus' response as a laudable gesture of humility, it was in fact much more.

Jesus credited his healing miracles to the faith of those who received them. By doing so he brought us face-to-face with a fundamental truth about the nature of reality: we are co-creators of our own destiny.

The Hindu concept of *maya* provides further insight into the nature of the co-creation process. Whether understood in its original sense as the creative power of God or in its more contemporary usage as the "illusion" under which we labor when we think of the world as a fixed and immutable solid, the term *maya* aptly describes the inherently fuzzy and malleable nature of the reality in which we live.

Fulfilling your role in the co-creation process requires more than merely giving intellectual assent to the basic principles involved. It requires a state of *enlightenment*. The dynamic and personal understanding of the unity of creation and our role as participants in the co-creation process has traditionally been won by the direct intervention of a master or through years of disciplined prayer, study and meditation, but our species now stands poised at the brink of a new era. Hyperspeed recovery happens at the intersection of personal intuition and quantum consciousness. If that's a neighborhood you haven't visited in a while, finding your way back at light speed may call for drastic measures. In the quantum realm things don't inch from one place to the next — they leap.

YOUR HEALTH,
YOUR DESTINY,
YOUR CHOICE

Y ou arrived in the world with fully functioning instincts and a keen sense of intuition.

What happened to them? Where did they go?

Nearly everything in your life is a byproduct of conditioning. From your earliest days, your thoughts, activities and preferences have been shaped by parents, siblings, teachers, friends, colleagues—and even the ever-present media.

As an infant, you knew perfectly well when you were hungry, sleepy or content, and to the best of your ability, you did something about it. At that tender age you weren't yet capable of meeting your own needs, so you relied on others for your survival. Those who cared for you did so in the way *they* thought best; in time, you came to perceive their ways of doing things as your own and stopped listening for the voice of your own inner wisdom.

You spent years in school being told to sit down and be quiet. Now that you've mastered the art of doing just that, everyone complains that you are an apathetic couch potato.

You were forced to ignore your body's natural instincts and instead learned to eat what and when you were told. Is it surprising, then, that you're so vulnerable to food ads or the latest diet craze?

Unnatural stool retention leads to constipation, abdominal cramping, diverticulosis and even colon cancer; nonetheless, you have been taught to ignore your body's natural and healthy urges for timely bowel movements in favor of a more "convenient" artificial schedule. Despite water's essential role in the human body, you have also learned to restrict your water intake so you can attend classes, business meetings and social events uninterrupted.

The modern world no longer resembles the natural environment you were designed to live in, so you've been forced to adapt to the artificial world we've superimposed on nature. Your body knows the difference. Busy schedules, clamoring alarm clocks and constant exposure to artificial lighting have shifted your circadian rhythms and left you chronically sleep deprived and dependent on caffeine and fast food to get through your days.

A lifetime of social conditioning has distanced you from your instincts and funneled

your lifestyle into some very unhealthy channels. Is it surprising, then, that we sometimes get sick? Really sick?

## WHO'S IN CHARGE HERE?

"These statements have not been evaluated by the Food and Drug Administration. This product is not intended to diagnose, treat, cure or prevent any disease. Please consult with your own physician or health care specialist before using this product."

Sound familiar?

As a species, we've been historically conditioned to venerate experts. As individuals, we've been trained from birth to do as we're told. As citizens, we've been the target of a fifty-year public relations and lobbying campaign designed to persuade us that when it comes to our health, the only experts we can trust are card-carrying members of the medical-industrial complex.

The fact is, conventional medicine does a spectacular job of managing acute health care problems like trauma and infections, but it has failed miserably in the arena of preventative care and the management of chronic illness.

The combination of the two—our unquestioning respect for conventional medicine and conventional medicine's limited range of capabilities—has been a prescription for disaster. When we get sick, a lifetime of conditioning tells us to turn to our medical authorities for solutions and answers. If the experts we turn to have no solutions to offer, we mistakenly assume that no solutions exist. We accept what we've been told and try to "make the best of things." When we take this approach, our quest for healing ends before it even begins.

I hope you don't find yourself in that situation. If you do, this is your wake-up call.

The human mind has always been at the forefront of medicine and healing action. We can take pride in twentieth-century medicine's success in reducing the risk of many infectious diseases. We should be grateful too for other advances in modern medicine, like the development of antibiotics, anesthesia and advanced surgical techniques. But if we as patients let the pageantry and gadgetry of modern medicine distract us from performing our primary role as healers, the entire process slows and eventually grinds to a halt.

The effects of this were painfully obvious in a clinical setting. Patients who were offered a broad range of care options and "followed their heart" in tailoring a personal wellness strategy achieved faster healing times at substantially reduced costs. Reconnecting with instinct and reclaiming a sense of personal autonomy allowed these patients to tap into the body's limitless healing potential. Providing my patients with full access to a broad range of complementary health care options allowed them to move beyond pain and disability into new dimensions of peace, confidence and unconditional happiness—fertile ground for recovery.

In contrast, those patients who didn't follow their inner instincts and instead adhered to narrow, preestablished medical and surgical guidelines deteriorated quickly.

When it comes to matters of your health, your goal should be to familiarize yourself with as many options as possible, so by all means, do consult with others—including doctors and the Food and Drug Administration. The right doctor can be an important advisor and gatekeeper—someone who holds the key to effective healing strategies you would not have access to on your own. If you are willing to do your homework and ask the right questions, the right doctor can help you understand unfamiliar medical terminology and research. The right doctor will be both your personal cheerleader and a key member of your health care team—but not the team captain. That's your job.

As you begin honing the instincts you were born with, they will help you tailor a wellness strategy that is just right for you. Because the fact is, when it comes to your body, *you are the expert*. You have lived with your body longer than anyone. You know its quirky habits best. And when the wonderfully complex body you live in breaks down, your return to wellness begins with the rediscovery of your own innate power of healing.

## CHILDHOOD 101

How many times did you skin your knees when you were a child? I'll bet you've lost count. You've recovered from more childhood illnesses and traumas than you can even remember. You did it all without thinking too, because your body came hardwired to heal.

Your instinctive cellular wisdom hasn't abandoned you. It's still there—right where you left it when you grew up and stopped listening to it. It's a little dusty, perhaps, but that can be fixed.

Do you remember how different life was when you were a child?

You felt your emotions deeply, expressed them without hesitation or apology and made no effort whatsoever to bury them or rationalize them away. When life pleased you, you giggled until it hurt. If life was treating you badly, you had a good sob until you got it out of your system.

Kudos. It was the right thing to do. It still is. Expressing feelings is a powerful means of keeping us in the moment—and that's where the healing happens.

Your emotions are your body's rapid response system for events taking place in your immediate environment *right now*. In fact, newly emerging evidence suggests that the term "rapid response" just might be an understatement: researchers have recently established a consistent pattern of measurable physiological responses in the human body *approximately five seconds before an unknown event occurs*. And while these pre-event physical changes can be detected in both the brain and the heart, they seem to show up first in the heart.[5]

While further research in this arena promises to give us a better understanding of how our bodies interact with time-space and the quantum realm, for the time being it is important for us to at least recognize our emotions for what they are. If we are perpetually sending out "feelers" into the

---

[5]McCraty R, Atkinson M, Bradley RT. "Electrophysiological Evidence of Intuition: Part 2. A System-Wide Process?" *Journal of Alternative and Complementary Medicine* 2004; 10:325-336

environment as a sensing mechanism, then burying the emotional reactions we have in response to a current situation makes no sense at all.

Feelings that occur midday in response to a particular situation are providing you with information that is relevant at that moment. The wrong thing to do is stuff those feelings way down inside until they break out in a stomach-knotting rant at the dinner table or a long, sleepless night of tossing and turning.

Putting your feelings aside until later is like reading last week's movie listings this week. All of the valuable information you could be receiving from them is lost because by the time you get around to processing those signals, they're no longer relevant.

When you do finally take time to deal with "old" feelings later, you again fall out of synch with the real-time events taking place in your life. Your emotional circuits are too busy processing old news to bring you up to date on information that would be of use to you *right now*.

Worst of all, the emotions we tend to postpone and rehash the longest are the ones with the most catastrophic effects on our bodies.

Now that we're adults, we have advantages we did not enjoy as children. We still feel our emotions every bit as deeply and immediately as we did when we were small, but we've gained the maturity and experience to respond appropriately to any situation with resources drawn from our own inner wisdom.

The right thing to do with emotional input is to pause briefly, close your eyes, and take a slow, deep breath. For minor events, the process may take only a few seconds. For major events, your best approach may be a five minute time-out for a quick round of the Breath of Fire—a powerful centering technique can help you find the voice of your own intuition.

Making a habit of allowing heart and mind, thought and feeling to achieve a harmonious "go" signal before acting is the key to inner wisdom and a happy, healthy, fulfilling life.

*Siddhartha Gautama (563? – 483? BC), better known today as the Buddha — The Enlightened One.*

## KNOW THYSELF

If you are new to the path of self-discovery, you may find that the process of inner listening doesn't always provide the clarity you need. At times you may feel like a cork being tossed about on a sea of conflicting voices and feelings. This is not surprising. You've spent so much of your life immersed in the thoughts, feelings and opinions of others

that you can no longer distinguish them from your own.

The real you is still there, buried beneath all those layers. You were meant to play the starring role in your own life — to be an actor rather than a reactor. Being yourself is the only qualification required for the job, and fortunately, finding your true self is not as complicated as it seems. All it takes is courage and a willingness to change.

We humans are creatures of habit. Much as we take the same route to work each day or eat over and over again at the same restaurant, we habitually engage in long-term relationships with just about everything in our lives. Over time, this sort of continual association blurs the boundary between self and the person or thing to which we've grown accustomed. We lose sight of who we are and think of ourselves only in terms of our relationships: we're a mother, a daughter, an architect, a department manager, a resident of the Willowbrook community, a golfer — but these titles describe what we do, not who we are. Getting to know your *self* again will allow you to tap into the energy source at the core of your being that was designed to provide healing and guidance *just for you.*

Meditation, yoga and Qi Gong are all wonderful ways of accessing the self, but when you are facing a terminal or chronic illness, stronger measures are called for. Awareness and self-discovery happen at hyperspeed when you shed habitual ways of relating to things and people and explore what remains: you.

## DETACHMENT

Two thousand five hundred years ago, the Buddha identified our attachment to things we can lose—things like money, career, home and relationships—as the cause of all suffering. The Buddha taught that this kind of attachment can distract us from discovering who we are and finding our true purpose in life.

I didn't understand the Buddha's teaching on detachment until I landed in prison. It was only after I was involuntarily separated from

> I didn't understand the Buddha's teaching on detachment until I landed in prison.

my wife, my children, my patients and my career that the depth of the Buddha's wisdom became clear to me. Prison taught me that I could be deprived of everything and still find contentment. It was a lesson I had struggled with all of my life; what I failed to achieve on my own journey of spiritual growth was accomplished for me by the U.S. government. I will be forever indebted to the U.S. Department of Justice, the F.B.I., and the Bureau of Prisons for revealing my path toward detachment and light.

The atoms and electrons that make up the human body are no strangers to detachment either. As you read these words, the atoms within your body are exchanging energy and information with thousands of neighboring atoms. Atoms cling to atoms but never indefinitely; these encounters are momentary and short-lived. Detachment from earthly distractions is easier when we release feelings of entitlement or possession and instead learn to think and act more like our subatomic inner selves. When you engage life without the boundaries and limitations of long-term attachment, you enjoy moment-to-moment pleasures in ways you have never appreciated.

When detachment comes involuntarily through the loss of health or a career or a loved one, it can be very painful, but even then it is not without rewards. No matter how much you lose, you never lose yourself. The emptiness you experience with loss can help you rediscover long-obscured territories of your own soul. When the initial pain of loss subsides, the once-forgotten vastness of who you are offers breathtaking rewards and unlimited potential.

When you experience detachment voluntarily, it's a much less painful process of stepping away from what has always been to take a fresh look at yourself from a new perspective. Since your habitual thoughts and actions are linked to your health, changing patterns can do more than yield a new, clearer perspective on yourself—it can spark dramatic and instantaneous change in your well-being. If your habitual way of life is contributing to your illness, then change is just the

medicine you need.

There are many kinds and degrees of detachment, each offering a unique path to enlightenment and self-awareness. Any of these time-tested methods can give you a fresh perspective and a new start.

## Detach from Your Old Beliefs

You won't be able to learn from new thoughts, ideas and dreams if you continue to cling to conflicting old ones. Every once in a while you have to empty your cup to make room for new sustenance.

Your body has the capacity for infinite storage, but like upgrading a computer, it is important to delete older files and programs to make room for new software. This is a vital step in your transformational process. To access new ideas and belief systems that are more pertinent to your new life, you may have to leave the comfort zone of your home, your career or even your established identity.

Become comfortable with your new direction in life and go it alone for a while until your old companions are ready to catch up with you. Remember, courageous men and women like you—those who dared to think the unthinkable and imagine the unimaginable—have always made the greatest discoveries in life. Congratulations for taking the first step.

### Detach from Your Past

If you wake up each day and remember a painful past, you are strengthening and re-creating the very events you find so painful. Is that really what you want?

One of the most astonishing findings of quantum physics is the discovery that our present observations have the power to change past events. Because the power you have to impact events in your life is available to you only in the present moment, the most self-empowering, life-affirming decision you can make is to abandon the past and live *right now* with a clear focus on creating your preferred future.

I will be forever grateful to the unknown inmate who alerted me to the deadly burden of anger I carried with me into the prison camp. Without his keen observation, I would never have noticed that every time I recounted the details of my legal case to a new audience, I was reliving painful memories and digging myself deeper into an abyss of bitterness and despair.

It wasn't easy to divest myself of the anguish of loss, suffering and legal entanglements while sitting in a prison cell. Frankly, my only contribution to the process was an initial commitment to stop talking about my past. From that moment onward, everything began to fall into place. Once my present no longer revolved around my past, I was again free to drink deeply of life and savor its many pleasures. With my attention turned away from myself, I became more observant and began noticing subtle beauty even in the austere environment of the prison. I established new friendships with highly intelligent and accomplished individuals, many of whom had stories more interesting than my own. In the weeks that followed, healing came naturally. I slept better, had fewer nightmares and noticed a significant decrease in heart palpitations.

I should not have been surprised. Taking a "vow of silence" as a first step toward recovery was a time-honored practice among my patients. I coached hundreds of individuals—beginning with my own father—to stop discussing their diseases. During my disability, my father fell ill and was admitted to the hospital, also in acute congestive failure. When we took a vow of silence and agreed that we would not discuss our disease with others, we both noticed dramatic improvements in our health. Banishing illness from everyday conversation is the first step toward allowing the body to believe everything is OK—even when it isn't. This is wellness conditioning at its best. This unique healing strategy worked for every patient who tried it and consistently followed through.

### Detach from the Ordinary

Travel is without question the most delightful way to experience detachment and gain a fresh perspective on yourself and on your world. If we live in the same cultural atmosphere long enough, we forget that cultural environment plays a role in our development, but it is not who we are. Losing sight of that distinction can separate us from important aspects of ourselves.

The remedy is simple: switch cultures, and be bold about it! The more cultural distance you put between your home country and your travel destination, the better. The sharp contrast will allow you to experience life anew, moment by moment—just as you did when you were a child.

If you aren't comfortable traveling alone, look for a new travel partner. Travel with a spouse or business associate makes it easy to slip into familiar patterns of thought and behavior. Traveling with a new companion will take you out of your comfort zone and make you think, feel and act in new ways. It will keep you in the moment.

Give yourself permission to break out of your shell and reinvent yourself as often as you can. When you immerse yourself in new sights, sounds, customs, people and foods, your body begins reinventing itself to better adapt to its new surroundings. Travel can unleash torrents of growth—a real advantage if you're stuck in a physical, mental or emotional state you'd like to change.

Even if you can't travel a world away, travel. Take a day trip to the next town over. Find a reason to spend the day on the road discovering your town and yourself. No matter what your destination, a glimmering jet slicing through a canopy of silver clouds can be your personal chariot to magical new dimensions.

The everyday world looks very different from thirty thousand feet. We gasp at the breathtaking beauty of snowcapped peaks, meandering river valleys and the verdant patchwork of farmland that seem so ordinary from the ground. But it's not the earth that's changed—it's us.

> Even if you can't travel a world away, travel.

The experience of being skyborne offers us the gift of a new perspective. For a few moments, at least, we see the world through the creator's eyes and catch a glimpse of immortality.

I feel gratitude, joy and hope each time I fly. I yearn to extend the sense of purpose and connectedness flight offers to every moment of my life. Each journey aloft is a divine gift, a heartwarming reminder of the essential unity we all share—one world, one family, one body, one God.

Travel offers more than a new destination.

It offers a new way of being.

### Detach from Stagnation

Whether you are a bedridden invalid, a housewife busy with toddlers or an executive working sixty hours a week, take time to broaden your horizons. Your mind and body thrive on challenge, novelty and adventure. Giving them what they need will pay dividends now and for years to come. No matter how busy you are, set aside time to pursue an interest and take on new challenges.

*Wellness at Warp Speed*

Devoting time and energy to your personal interests is an essential part of being alive. If you always dreamed of being an artist, start. Paint or join a pottery studio. If you love to cook, ratchet your culinary skills to the next level with a class on exotic cuisines or lavish desserts.

Your interests are life's way of reminding you that you came here for a reason. If you pursue your dream, no matter how silly or far-fetched, you will tap into a rich source of growth, purpose and fulfillment that will bring sparkle to your life when everything else fails.

Pursue your dreams, but don't stop there. Challenge yourself in unfamiliar arenas. If you sit at a desk all day, try jazz dancing or skydiving. If you're engaged in a physical occupation, tackle a new language or join a poetry circle and begin expressing your creative side. Take a class in glass blowing, coach little league or learn to play an instrument.

*Helen Keller (1880 – 1968)*

Taking on a new challenge will do more than foster the development of your brain's neural network. Exploring the unfamiliar can lead to the discovery of unexpected delights that will enhance your life for years to come. Most important, whether you succeed or fail at your new endeavor, the simple act of taking on a challenge will teach you something important about yourself: you are stronger than you think you are.

Whether you've lost your job, fallen into a ravine or have just been diagnosed with a life-threatening illness, the confidence that comes from knowing how to set aside your fears and tackle an unfamiliar situation can save your life.

### Detach from "Security"

If you love your work—if you wake up in the morning eager to begin another exciting day on the job—then feel free to skip this section.

If, however, you drag yourself to work each day at a job you hate for the sake of a paycheck, you don't need me to tell you it's time for a change. We all know that job-related stress contributes to millions of untimely deaths each year. Have you made it your personal goal to add to these statistics?

If not, my question to you is this: What are you doing about it?

Most people who feel trapped in their jobs aren't really nailed to their desks—they are confined by habit and the sense of security it provides. If you are healthy enough to be working, you can probably list a dozen reasons to postpone your job change.

That's the problem: you are too healthy. Knowing that you don't have to make an immediate change keeps you in the wrong job and prevents you from finding the right one. If this describes your situation, you have two options. You can stay where you are until events beyond your control force you out, or you can make a firm commitment to improve your situation and exercise the discipline and will to carry it out. The choice is yours.

If you enjoy your work but not your job, update your résumé and search for a healthier way to use your skills and experience. If you're past due for a career change, pursue training in a new field or start your own business.

If you have no background in a career that's of special interest to you, terrific! Find a way to volunteer in the field in your spare time. Even unpaid work will give you valuable work-related experience and allow you to develop a network of supporters who can help you land your dream job.

Above all, don't succumb to the belief that you are stuck where you are—or worse, that you will make a change later. You won't.

> Security is mostly a superstition. It does not exist in nature. . . Life is either a daring adventure or nothing.
>
> – Helen Keller

### Detach from Stifling Environments

Is it time for a change—of address?

By the time we're adults, most of us can identify symptoms that signal the onset of illness in our own bodies. We recognize the sneezing, itchy eyes and runny nose that signal the onset of allergy season, or the queasiness and light show that precede a migraine. What we don't realize is that disease itself is symptomatic of a life that's run amuck.

Deteriorating health triggers a round of second-guessing. When things go wrong, we critique our eating habits and wish we exercised more—or less. We wish we had quit smoking sooner. We wonder if more sleep or less alcohol would have kept us alert behind the wheel.

Self-condemnation is pointless and unhealthy, but the regrets and questions that arise from self-examination can identify key strategies to help us recover. Those who experience "impossible" recoveries understand that a serious illness is not a death sentence—it's a call for change.

That change begins with your decision to make recovery your first priority. Your commitment to change is the first step on a journey that involves a complete transformation of your thought patterns and lifestyle.

Most of us habitually resist change because we think of it as an unpleasant disruption in the carefully constructed order of our lives. When change comes in unexpected and involuntary ways, it involves the loss of something we value. It's important to remember, though, that while this kind of change does occur, it's only half of the equation. The changes we find difficult and threatening are the ones that happen without our consent. When events beyond our control leave us struggling to find our bearings in an unfamiliar situation, it's not surprising that we find the entire process intimidating.

Under other circumstances, change isn't a bad thing at all—it's a great way out of a bad situation. The changes you make voluntarily won't diminish your sense of control—they'll enhance it. Taking steps toward a destiny you've chosen for yourself is an act of personal empowerment.

A serious or lingering illness may provide

> Synchronicity lifts the veil of separation between the physical world and the collective field of consciousness that holds everything together.

us with the incentive to overcome our own resistance to change, but it will not necessarily have an impact on the lives of those around us. You may be willing to quit drinking to save your liver; your drinking buddies may not.

A greater hindrance may come from well-intended people in your life who aren't convinced that you can recover. This can present a special challenge. Some of the people closest to you may express their heartfelt best wishes but consciously or unconsciously deplete your energy and undermine your wellness strategy with their own doubts and fears. Don't let them. Right now, your needs come first—not their feelings.

A change of location can help you set aside the demands and expectations of others and focus on your own needs. Your condition may benefit from a change in climate, from medical resources not available in your community or simply from a more life-enhancing environment. A move across town or across the world will allow you make the choices that are right for you and focus your energy on a joyful recovery rather than on justifying your decisions to others.

The decision to move is an intensely personal one. Only you can make it. In my case, it was the right thing to do. My decision to abandon my deathbed for a sunny beach in Mexico defied medical logic. It meant leaving behind those who loved me but doubted my ability to recover. My instinct-driven leap of faith took courage and steely determination, but it set me on a path of miracles that led to my amazing recovery from a deadly heart condition.

What about you? Who belongs on your recovery team? Who might unwittingly sabotage your chances for success? If you got sick where you are now, is it so illogical to think that you might get well somewhere else?

### Detach from Unhealthy Relationships

A serious illness calls for the reevaluation of every aspect of your life—including your most significant relationships. Booking a trip to Fiji may be easier and more fun than walking away from a ten-year relationship with a primary partner, but if that relationship is inhibiting your ability to recover, you should probably make that trip a solo flight—one way.

Each new relationship begins with the hope that we've found a partner who will be a long-term source of strength, nurturing and companionship, but things don't always turn out that way. People grow. People change. And sometimes people just plain turn out to be "not as advertised." What then?

There are no perfect relationships. Even the best of them may involve give-and-take cycles that even out over a period of years rather than days or weeks. In most cases, overcoming the challenges involved with establishing and maintaining a healthy relationship can become a source of growth for both partners. Over time, the process of working through life's challenges together—including the challenge of relating to each other—builds a shared history of success that becomes a source of pride, strength and confidence for both partners.

Unfortunately, some relationships are unhealthy to the point of toxicity. If you're in one, you are in the wrong place. Living with the constant stress of an unhealthy relationship will diminish your chance of recovery now and leave you more vulnerable to illness in the future. If a healthier, happier destiny awaits you elsewhere, why stay?

### Detach from Identity

What's in a name?

A lot. In recent generations, most Westerners have lost sight of the fact that names have meanings. The surnames we inherit usually describe the occupation, characteristics or place of origin of a distant ancestor. Daniel Cooper's early progenitors made barrels for a living. Zoe Chandler's ancestors stocked the ships of early seafarers with candles and other necessary supplies. Sandra Madrid's family hailed from the heart of Spain.

Our names are our most intimate and personal description of who we are as individuals, but most of us think of our names as nothing more than a familiar set of syllables. Our ancestors labeled themselves and one another with titles that described their lives in a particular time and place. We in subsequent generations have retained the original titles, but we've lost sight of the relevance and meaning those titles conveyed.

The ancient Jewish scriptures speak of a time when Abram, whose name meant exalted father, was given a new name: Abraham, the father of many. The name change, given by God, was a fitting recognition of a transformation that was about to take place in his life. The man who would one day be revered as the father of the Jewish people would also sire other great nations through his son Ishmael.

While many ancient traditions recognized the value of changing one's name to mark an important life transition, few traces of the practice remain in the modern world. Women may adopt a new surname at marriage and by doing so identify themselves in terms of a new relationship. Divorce and remarriage can lead to a series of changes. Those hoping to "make a name" in the entertainment industry choose a distinctive new moniker with broad public appeal. Authors adopt pen names not only in hopes of enhancing the marketability of their work but to establish a literary identity that fosters a personal sense of creative freedom.

Your name defines who you are. It shapes your image in your own mind and in the minds of others. What does your name mean to you? Does it reflect your life's purpose? Does it inspire you? Or is it just a meaningless set of syllables you've long since outgrown?

My own birth name—Nasser Talebzadeh Ordoubadi—was bestowed upon me by my father in honor of Gamal Abdel Nasser, an army officer who rose to prominence on the world stage in 1954 by deposing Egypt's King Farouk. In 1956, Nasser moved to nationalize the Suez Canal, an act that sparked the international Sinai Crisis. In the year of my birth, the French, British and Israeli armies joined forces to vanquish Nasser's army, but Nasser parlayed his defeat into a full-scale political victory and was soon recognized as the leading spokesman for the region's underdogs—Arabs and the developing nations of Asia.

My father, a free-thinking Iranian general in the shah's army, had had his own run-ins with the inefficiencies of a power-hungry, super-centralized regime. Known as "the McArthur

*Wellness at Warp Speed*

of Iran," he valued independence of thought and action above all else. Naming his son after a military officer who reached beyond the boundaries of convention and defeat to become an important figure on the world stage was an act much in keeping with his character.

As a young American doctor working hard to establish my first practice, I quickly realized that Nasser Talebzadeh Ordoubadi was an unsuitable name for a professional working in the United States. Patients and colleagues found my name difficult to pronounce, hard to remember and nearly impossible to spell. In 1987, I approached my father with a purely pragmatic suggestion of a name change. The idea met with a cool reception. In my father's eyes, generations of proud history were riding on my name. Out of respect for my father's wishes, I decided to postpone my decision.

Disaster brings out the best and worst in people. In the wake of the attack on the World Trade Center, America was flooded with an outpouring of grief and sympathy from around the globe. The enormity of the devastation touched hearts everywhere; for one stunned,

grief-stricken moment, mankind stood united in an instinctive realization that something was terribly wrong on our planet.

That brief moment of unity did not last. It was followed immediately by the polarization of nations, leaders and individuals. While some saw in the tragedy an opportunity to create a more loving, just and compassionate world, others succumbed to blind rage and a seething desire for vengeance. The drama played out not only on foreign shores but on the streets of America, where having a name of Middle Eastern origin became a dangerous liability overnight. Debates about the ethics and efficacy of racial profiling as a security measure obscured the fact that a more covert kind of racism was taking place behind the scenes. Innocent individuals and corporations with "wrong"-sounding names lost already-negotiated contracts overnight. Ordinary men and women were denied the opportunity to open bank accounts.

The persecution didn't stop with economics. When innocent Americans fell victim to misguided acts of retaliatory violence in the weeks following 9/11, I'd had enough. I changed my name to Noah A. McKay the following summer.

My new name was chosen to resonate with my personality, destiny and present direction in life. Noah predates the longstanding rivalry between Jews, Christians and Muslims, underscoring my commitment to political and religious neutrality. My surname was chosen in honor of my wife's grandfather, a humble, hardworking and well-liked California ranger who faced challenges of his own as one of an earlier generation of Irish immigrants.

While my birth name reflected the values of a man I loved, respected and admired, my chosen name stands as a living affirmation of who I am today. Each time I hear my new name, I am reminded of my own values and purpose in life. Each signature affirms my commitment to act in accordance with my own deeply held principles and beliefs. It's been a wonderful experience.

A name change should not be undertaken lightly—it's a process to be savored. It's a fun process that will allow you to step away from a past someone else has chosen for you and carve an important toehold in a future of your own making. Try on a few new identities. Decide for yourself what fits and what doesn't. The process of selecting a new name will give you an unparalleled opportunity to discover who you are now. Making your name change official will establish a real-world birth date for the person you'd like to become.

## THE HIDDEN POWER OF CHANGE

Some of the reasons to embrace lifestyle changes are obvious. It requires no great stretch of logic to understand why it's a good idea to quit smoking, switch to organic foods, or reduce your overall exposure to synthetic materials and electromagnetic fields. What may be less obvious are the benefits we receive from embracing change for its own sake.

Our ability to influence the unfolding of events in our lives exists only in the present moment. When we habitually repeat the same behaviors over and over again, they eventually become so engrained that we engage in these activities—things like driving, eating and shaving—without thinking about them. Our thoughts are elsewhere—or else*when*.

Children, on the other hand, remain intensely engaged and aware at all times because every moment brings a new adventure. For children, the simple act of toothbrushing is a rich sensory experience filled with interesting tastes, sensations, sounds and splatters. In time, brushing becomes such a routine part of our grooming ritual that we hardly notice it. Our bodies engage in the act of brushing, but we don't. Our thoughts are off mindlessly rehashing the events of the day or wrestling with an overbooked "to do" list.

As we age, an increasing percentage of our behaviors become habitual, and we spend more and more of each day disconnected from the present and from ourselves. In these disconnected states, we deprive ourselves of the power of the present moment and lose the cohesive state of physical, mental and spiritual awareness that allows us to take advantage of it.

Change restores awareness, cohesiveness and the power of the present. Don't believe me? Rearrange your furniture. When you walk into the living room tomorrow morning, you will *see* your environment in ways you haven't for a long time. You will be *present*.

The importance of being present is an ancient theme handed down to us by each of the world's great religions, but its role in healing has so far gone largely unrecognized by Western medicine. This is about to change. The discoveries of quantum physics have given us the science we need to bridge the gap between spiritual practice and practical medicine.

The most tragic victim of the Copernican revolution was not Galileo — it was mankind's pursuit of knowledge. Despite the fact that anyone with access to a telescope could see that church authorities were wrong about the movement of planetary bodies, ecclesiastical scholars stubbornly adhered to centuries-old teachings based on little more than a traditional misinterpretation of a few verses of Holy Writ. This intellectual arrogance led to a scientific backlash; in the end, science won the day.

"Show me the evidence" became the new rallying cry of science. It would remain so for centuries to come.

The trouble is, the new scientific definition of "evidence" was limited to the tangible realm. Only those things that could be measured and studied with the rudimentary technologies of the day were deemed "real" and of scientific value. Everything else was dismissed as superstitious nonsense. Over the centuries, the philosophy of scientific materialism took root and grew. By the dawn of the twentieth century, few scientists could imagine anything else.

If we bear in mind that this was the philosophical backdrop against which early quantum pioneers like Niels Bohr carried out their work, we can understand why his oft-quoted remark "Anyone who is not shocked by quantum theory has not understood it" may be famous, but it's not very accurate. Scientists may have been shocked to find their long-held assumptions about the nature of the universe crumbling beneath their feet, but frankly, they were about the only ones.

Every spiritual tradition in the world traces its roots to founders who grasped something about the fundamental nature of reality that most people had overlooked. That understanding allowed them to walk between the worlds, to predict the future, heal the sick, raise the dead and change the physical properties of matter. Astonished witnesses may have seen these events as miraculous suspensions of natural law, but the spiritual masters knew that they were acting in perfect harmony with a more fundamental set of laws the onlookers did not yet understand.

Throughout the ages, mystics of every tradition have rediscovered the truths their masters sought to impart. Wherever these truths prevailed, miracles followed.

Fortunately, most of us do not live in the insular world of materialistic science. We live in the real world, where "miracles" seep through on a fairly regular basis.

You've probably experienced a few yourself. Nearly everyone has a family ghost story, an episode of precognition, an object that inexplicably disappeared—or reappeared, seemingly from nowhere. We just don't talk about these things. We're afraid people will think we're crazy.

You're not. You've just been catching an occasional glimpse of the *real* reality. The big one. The one your spiritual tradition's founders were trying to show you.

One of Carl Jung's most significant contributions to the field of psychology was his definition of the concept of synchronicity. Synchronicities are meaningful coincidences so unlikely that they cannot be explained by chance alone. When two people on different continents call within hours to suggest that you read the same obscure book, you can't help but wonder, *"What was that all about?"*

That's synchronicity.

Events like these are signposts. Watch for them. If synchronicity is playing a greater role in your life these days, you needn't worry that you are losing your mind. On the contrary! You are growing more adept at tapping into indivisible nature of the quantum universe.

Your miracle is out there. Go find it.

You've caught a glimpse of what physicist David Bohm called "the implicate order."[6] Synchronicities are momentary fissures that allow us a brief look into the limitless, undivided universe underlying every aspect of nature. Synchronicity lifts the veil of separation between the physical world and the collective field of consciousness that holds everything together. We think of our world as a fragmentary place where everything is separate, solid and unchanging, but the true nature of reality is much more fluid. Reality is built of energy shaped to conform to the heartfelt beliefs and expectations of the observer. Throughout history, the enlightened individuals who understood this principle used it to create miracles. Their message to you is this: you can too!

If you achieve nothing more than an academic understanding of the principles of quantum physics, you will remain stranded in an intellectual house of mirrors. Your everyday

[6]Bohm, David, *Wholeness and the Implicate Order* (New York, Routledge & Kegan Paul, 1980)

world will appear fixed, solid and separate. Quantum science will tell you that it's not. Carrying around two such divergent sets of beliefs about the nature of reality will leave you as befuddled as poor Niels Bohr.

The power of the quantum realm is available to you only in states of coherence. Some call this being *in the zone*; others call it being *in the flow*. When heart and mind are fully present and in a harmonious and unforced state of agreement, anything is possible.

Hearts do not speak in a language of words. They operate in a realm of feelings generated by experience. Honor your moments of synchronicity as messages from the quantum realm. If you do, more will come. When you experience a moment of temporal distortion—if events happening around you seem to speed up or slow down—take it as reminder that your relationship to the time-space continuum is flexible, not fixed. If you worry about your daughter all morning and later discover that she's gone into labor early, you can be certain that the connection between you is more than imaginary.

Synchronous and mystical events, no matter how small, provide your heart with the growing body of experiential evidence it needs to align itself with your intellectual understanding of spiritual teachings and quantum principles. Synchronicities are happening in your life right now. If you haven't noticed them, it's probably because you've been blindfolded by habit.

There is tremendous power in learning to see, feel and experience life anew. That power comes from change. As your heart and mind open to the challenge and excitement of new possibilities, your body will begin gearing up to take on whatever comes next. Change brings a sense of freshness to life that will keep you on your toes and in the moment. That's a good thing: it will help you rediscover your *self* and reclaim your rightful place in the cosmic field of consciousness, where creativity and potential are limited only by your imagination.

Your miracle is out there. Go find it.

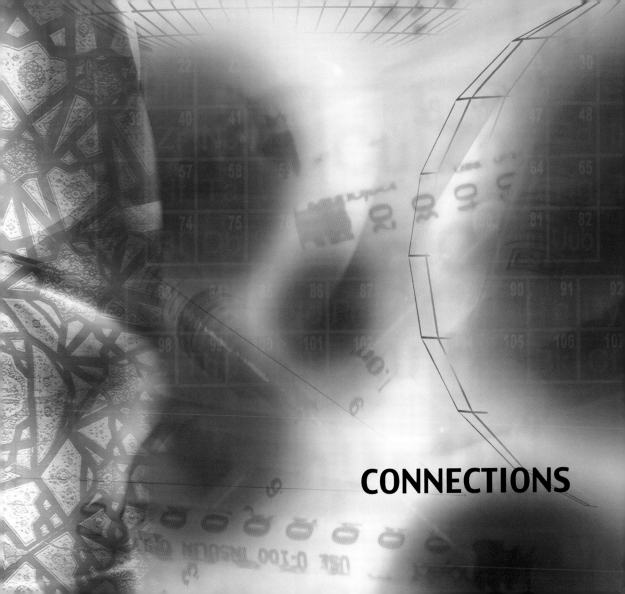

CONNECTIONS

By now you've noticed that this is not your usual health book. We haven't talked about the importance of calcium supplementation and veggie servings or the evils of white bread and transfats. You already know that smoking is a bad idea, and as far as I'm concerned, whether you get your daily aerobic activity in two brisk fifteen-minute walks or one long run on a treadmill is entirely up to you. You don't need me to remind you that lifestyle, diet and exercise play an important role in maintaining a healthy, happy life. You know that already.

What you may not know is that when it comes to your health, some things are more important than antioxidants and cholesterol levels. During my twenty-year career as a practicing physician, I cared for thousands of patients who were experimenting with different models in wellness. Like all practicing physicians, I had patients who broke every health rule in the book and still enjoyed long, healthy, happy lives. Conversely, patients who are paragons of healthy living sometimes contract serious illnesses and die relatively young. This raises a question few in the medical community have been willing to confront: are the experiences of these patients meaningless anomalies—or do they represent an important but overlooked piece of the wellness puzzle?

My goal in writing this book has been to introduce you to the other kind of health plan—the one embraced by unsinkable patients like Louise, a fiery little redhead who recently celebrated her hundredth birthday. She lives independently in her own home and tends a thriving flower garden. She keeps a dizzying social calendar, watches over friends, family and strangers alike and enjoys a rich spiritual life.

What's Louise's secret? Well, let's see. She surrounds herself with light and color. She's a great dancer. She partied every weekend and ran hooch during Prohibition. She has a wonderful sense of humor and loves a good joke. The only vegetables she will eat are coffee and chocolate, both of which, she wryly notes, come from beans.

Does she worry about cancer? Nope.

Louise long ago decided that she would never be afraid of anything. Besides, when it comes to cancer, she's been there, done that. Heart disease? Been there too.

In fact, heart disease transformed Louise's life. Fifteen years ago, she lay in an emergency room listening to a cardiologist's gloomy prognosis: she needed surgery, but was too old and frail to survive it. He promised to make her as comfortable as possible but could do little more. He offered her some pills and advised her to just go home and enjoy whatever time she had left.

You don't live to be a hundred by ignoring your own intuition, so Louise asked for a little time alone. When the cardiologist left the room, she began to pray. Within minutes, she had a vision. A warm, golden glow filled the room, and she was suddenly surrounded by dozens of people. Some of them she knew to be living, others had long been dead. They had come to bring her a message: *your work here is not yet done*.

Louise was shown a world full of lonely people and made to understand that it was her job to reach out to them. She got the message loud and clear. When the cardiologist returned, she insisted that she would have the surgery immediately. Her determination left him a little flustered. He reiterated his belief that she would not survive the operation.

She looked at him with steely blue eyes and replied with just two words: *"Wanna bet?"*

The surgery went without a hitch. Louise was back home in no time and today plays an indispensable role in her community. She makes calls to check on elderly shut-ins, provides food and shelter for the homeless and never forgets a birthday. Louise pours a great deal of love into her world, and her world loves her back. She's an inspiration to everyone around her.

Something tells me she would have been fine with or without the operation. In the midst of illness, Louise's intuition led her to the infinite source of life. Her connection with God and the quantum field did not begin with her deathbed vision—they had always been there. They are inseparable from her very existence. The only thing that changed was her level of

personal awareness. No one will ever be able to persuade her that the spiritual realm does not exist. She knows better. She's found her divine purpose and understands her intimate level of connection with God as the unlimited source of power. The loving, healing energies that flow through Louise into the world around her will heal and sustain her until she's finished her appointed task.

## THE BUTTERFLY EFFECT

We've all had the unpleasant experience of walking in on an argument. Even if everyone present manufactures a smile, the tension hangs in the air like a shroud. Everyone can feel it, and no one likes the sensation.

Conversely, do you remember the little rush of pleasure and gratitude you experienced the last time someone stopped to let you out of a parking lot onto a busy street or invited you to step ahead of them in line at a grocery store? For one brief moment, a stranger held in his hand the power to shape your view of the

*Earth as seen from Apollo 17 — December 7, 1972.*

world — and shaped it for the better.

That stranger's tiny act of generosity prompted you to do the same for someone else. When you passed that little act of kindness on, you experienced another little rush of pleasure — and so did the recipient of your compassion.

Most of us think of our feelings as passive, involuntary responses to events taking place in our lives. They are much, much more. Emotions are contagious. They are the most powerful tool we have to influence the conditions, people and events shaping our daily lives.

Don't believe me? Try this experiment: for the next thirty days, generate love in your world. Lavish affection on your partner and children. Be kind and attentive to your pets. Treat friends, neighbors, co-workers and strangers like royalty. Look for simple ways to surprise and delight everyone around you. Pray for the well-being of everyone in your life — especially your enemies — every single day.

Don't expect anything in return. Don't tell anyone what you are doing. Just generate love and observe.

## THE BANDWIDTH OF LOVE

In order to thrive, we need a sense of purpose and a continuous flow of life-sustaining energy. That energy reaches us in many different ways. It comes in sunlight, in the nutrients present in fresh, healthy foods, and in the infrared energy we exchange during a hug.

You know that five minutes spent racing across a hot parking lot to reach the car before rush hour is not the equivalent of five minutes spent relaxing on a sunny beach. You know too that a quiet meal in a relaxed, friendly environment impacts your body very differently than the same meal choked down in the wake of a tense meeting. And you most certainly sense the qualitative difference between a formal, back-slapping business hug and one exchanging genuine affection with a loved one.

The sunlight is the same. The nutritional value of the food is the same. The hug's duration and degree of physical contact are the same.

What makes these experiences so different? Bandwidth. When you are relaxed and open, your bandwidth increases, allowing more life-giving energy to flow into and through your body during your energetic encounters.

There is a reason the world's great spiritual traditions have advocated love, forgiveness, mercy, generosity, gratitude and compassion. These emotional states all increase your spiritual energetic bandwidth. Others—like anger, bitterness, hatred, envy and fear—reduce it or shut it down completely.

Your emotions are not just internal sensations. They are the mechanism that governs the quality of your energetic exchanges. More important, as part of a two-way communications system, they ripple outward like waves on a pond and have an impact on everything and everyone around you. Sooner or later, they'll be reflected back to you in the form of your future.

What that future looks like is up to you.

## 9/11: THE WAKE-UP CALL

The first plane could have been anything. The second could have been nothing else.

The grim, slow-motion collapse of the World Trade Center's second tower on the morning of September 11, 2001, brought my life as a physician to an abrupt end. It changed your life too. From that moment on, nothing would ever be the same again, not for any of us.

I was preparing for work when the news broke—not in my role as CEO of one of the largest private medical practices in the state of Washington but as a bathroom orderly at a federal prison camp in Sheridan, Oregon. Barefoot, shivering and clad only in a pair of white prison-issue boxers, I rushed to join the dozens of inmates huddled around the camp's small television set.

A part of me shriveled and died that day with the souls trapped in the crumbling towers. I'd spent four of my most memorable years living and working in the Bronx as a medical student. The landmarks of Lower Manhattan were like a second home to me. Each slow-motion replay of the tragedy plunged me deeper into a black hole of horror and disbelief. How could anyone contemplate such an unspeakable act?

I felt an instinctive call to action. I wanted to scream out loud, loud enough for the whole world to hear, but my feelings were too raw and intense for words.

I was not alone in my stunned silence. The connection between the bleak images on the screen and the lonely, desperate eyes of those around me pierced my side like a hot dagger. For prisoners, 9/11 took place in a world intimately connected with our own and yet entirely beyond our reach. Like Americans everywhere, we stood transfixed by the images unfolding on the small screen, struggling to take in an event bigger than any of us could have imagined.

Silence hovered over the camp for hours before the words began to come. At first, each

*Disciples following in the footsteps of Jesus, played by Ahmad Solei-manniyeh and the cast of "Jesus, Spirit of God."*

inmate's thoughts were expressed in a style as unique as his identification number, but once the trickle of words began, they quickly coalesced into a great flood of sentiment ringing throughout the compound: "It's time for a change, man. It's time for a change."

If the world wasn't visibly polarized prior to 9/11, it most certainly is today. The perpetrators of the attack intended the event as a wake-up call for the millions who've lost sight of the social and economic realities of our time. For some, the World Trade Center stood as a proud symbol of America's hard work, achievement and economic success. For others, it represented a shameful history of greed, arrogance and global exploitation.

Two thousand nine hundred seventy-three people were caught in the middle.

No one talked about the perceptual divide before the attacks. Once they occurred, the emotional response driving early opinion polls was anger and a burning desire for retaliation. Battle lines were drawn before the dust had even settled, and before long, everyone felt compelled to form an opinion and take sides.

For better or for worse, the war was on. Six years later, each side is more deeply entrenched in its position than ever, and with greater conviction. The initial global outpouring of sympathy that held the potential to unite mankind in pursuit of a better world has been replaced by a worldwide sense of hostility, distrust and despair. The once-bridgeable divide between "them" and "us" has become a vast chasm that now threatens to swallow us all.

It's difficult to argue that the six years of deadly conflict that have taken the lives of more than three thousand American service-men and women and hundreds of thousands of Iraqi civilians have made us any safer. It's also obvious that we are no closer to containing and eliminating the causes of global terrorism—on the contrary, the bloodshed and massive destruction have only widened the conflict and created new breeding grounds for hate.

Two thousand years ago, Jesus issued a seemingly absurd commandment to love our enemies. Was he as crazy as his critics claimed—or did he understand principles the rest of us are only now beginning to figure out?

Now more than ever, what we need both in our personal lives and in our societies is more compassion, a greater depth of understanding, and especially more love. Love is a universal force, a quantum equalizer and the only real answer to the problems facing us today.

## COLLATERAL DAMAGE

President Franklin D. Roosevelt once inspired a depression-weary American populace with the unforgettable phrase "the only thing we have to fear is fear itself."

The philosophy touted by today's world leaders is somewhat less inspirational, and was best summed up in a line once uttered by the

Dr. Martin Luther King, Jr. at age 25. He was later awarded the Nobel Peace Prize for his successful leadership and advocacy of non-violence during the American civil rights movement.

121

young Wednesday Addams: "Be afraid. Be very afraid."

We've come a long way, but not necessarily in the right direction. We humans come wired with a sophisticated threat-response system designed to keep us alive in dangerous situations. The screech of tires warning of an impending car crash bypasses our body's usual information processing system and takes the fast track to a small, almond-shaped portion of the brain called the amygdala.

The perception of an immediate threat triggers a host of changes within our bodies. We freeze for an instant to make an unconscious assessment of our situation: Where is the danger coming from? Should I duck or run? In the fraction of a second it takes to size up the situation, adrenaline surges through the body, preparing it for action. Your pulse quickens, the physical characteristics of your blood change, your breathing becomes rapid and shallow and your blood pressure soars. Your body is gearing up

> Our scientific power has outrun our spiritual power. We have guided missiles and misguided men.
>
> – Dr. Martin Luther King Jr.

for battle—or flight.

We recognize this response as fear. Fear is a valuable ally when you are faced with an immediate and unexpected peril. It's a primitive but effective means of staying alive when there's no time to think. Allowing sensory information to bypass the reasoning parts of your brain gives you a slight reaction-time advantage. That fraction of a second may be just enough to allow you to swerve out of the path of an oncoming car.

But what happens if the tension never goes away? What if you spend weeks, months or even years stuck in a state of heightened fear? We in the medical community have a word for that kind of unremitting pressure. We call it stress, and it's a major contributing factor in every illness you can name.

Crisis-based emotions like anger and fear stem from a primitive survival response system that is hardwired into the neural circuitry of all living creatures. A mouse, a snake or even a tiny insect will attack if it feels threatened. We humans have not outgrown the survival response system we share with other creatures; it remains with us today as a part of our basic physiological make-up.

What makes our kind unique is our ability to use the reasoning power of more advanced states of consciousness to evaluate and over-ride our more primitive impulses.

In any given situation, we have the power to moderate and transform our response to one of neutrality or compassion. It's important that we do so, because the body's fear-based rapid-response system is only designed to cope with a brief and immediate threat. It won't factor in the long-term consequences of a possible course of action. It's not designed to make sound policy decisions. As long as we remain in a state of fear or anger, we limit our access to the brain's advanced reasoning skills and lose the wisdom, logic and analytical ability that make us human.

Fearful, angry leaders make for fearful, angry societies, so it's not surprising that things aren't going so well in the foreign-policy arena. The deadly consequences of living in a perpetual state of anxiety are well understood in the lives of individuals. What we are witnessing now is a tragic demonstration of the effect this kind of fear-based decision making has on a global scale.

It's awful to consider the possibility that we may be unwittingly driving up our own death toll by living in constant states of fear, anger and hostility. It's even worse to realize that these negative emotional states have dis-tracted us from coming to grips with an even deadlier peril.

## THE GREATEST THREAT

For more than a century, the world's public health officials have struggled to identify and eliminate harmful microorganisms, with no end to the battle in sight. I understand their plight. I spent much of my career searching for new weapons to use in the battle against deadly human pathogens.

But today we are under attack by an even greater menace, a stealth killer posing a more immediate threat to the survival of our species than any plague in history. The cause of this epidemic is invisible, ubiquitous and deadly. Tragically, the nature of the threat has gone unrecognized because the unimaginable scale of the devastation left in its wake has blinded us to the possibility that we are dealing with a single underlying cause.

The 1918 outbreak of Spanish influenza is considered the greatest pandemic of the twentieth century. This global outbreak killed a staggering *fifty million people*—one of the greatest death tolls for any disease in human history.

Despite the unimaginable suffering implied by those statistics, it is important for us to realize that a more deadly and insidious killer has gone largely unrecognized. In the twentieth century alone, two hundred million men, women and children died of a single, preventable cause: *violence!*

Shortly after the terrorist attacks of September 11, 2001, the prestigious World Health Organization released a comprehensive report on global violence based on three years of investigative research by 160 professionals and specialists from around the world.

The results? 1.6 million of us lose our lives to violence *each year*. The devastating effects of violence are not limited by race, gender, geographical region or social class. The plague of violence affects us all.

Many assume these deaths can be attributed to acts of war or terrorism, but nothing could be further from the truth.

**The Facts:**

• **1,600,000 people die from violence each year. If this year is typical, then:**
• **310,000 of us—20% of the total—will die as victims of war and genocide**
• **520,000 of us—30% of the total—will die of violence inflicted by a family member or friend**
• **815,000 of us—50% of the total—will die of suicide and self-harm**

Can you believe it? Our trusted friends and family members are more likely to kill us than terrorists or invading armies—and all of our angry friends and enemy combatants *combined* are less likely to kill us than we are to kill ourselves!

We are the greatest threat to our own survival!

## THE GOOD NEWS

These statistics may be shocking, but they're not really surprising. The world we live in is a mess. Decades of fumbling efforts to engage in better living through chemistry have left us swimming in a toxic soup of pollutants that has permeated every cell in our bodies. Generations of incoherent policy decisions on the part of the world's leaders have left us teetering on the brink of unprecedented-scale disaster.

We're living in consumption-driven economies where advertisers remind us each day that we are inferior and incomplete if

*Mahatma Gandhi (1869 – 1948)*

we don't own the latest products. Half of us are paying for "the good life" with high-stress jobs and lifestyles that will kill us early; the other half are trying desperately to reach the bottom rung of a ladder that ultimately leads to nowhere.

We take pills to cope with depression and anxiety. We take pills to help us sleep. We take lattes to help us wake up. Our bodies are bombarded with a constant stream of electromagnetic radiation from the high-tech devices we work so hard to buy.

But the most deadly threat we face is the loss of purpose and meaning. Why *do* the years we spend in the pursuit of material goods ultimately prove unsatisfying? How can a yogi with no possessions find perfect contentment meditating in a cave? How can one tiny Albanian nun named Mother Teresa take a vow of poverty and change the entire world?

How could Mahatma Gandhi's unyielding devotion to peaceful resistance and a simple, non-materialistic lifestyle drive out the powerful forces of the British Empire and establish the world's largest democracy? Why do our efforts to achieve the same ends with guns and bombs so often end in disaster?

Our earliest ancestors perceived the world through the eyes of instinct. Their intimate connection with the world around them allowed them to grasp fundamental principles about the nature of reality that mankind would later forget—until now.

We have come full circle. A century ago, science described the world as a rational, predictable, non-spiritual universe composed entirely of matter acted upon by impartial forces. That early scientific obsession with matter spawned a broader societal obsession with materialism that has had devastating social and environmental consequences for us all.

> Victory attained by violence is tantamount to defeat, for it is momentary.
>
> – Mahatma Gandhi

If there is hope to be found in our situation, it is in this: things are about to change. The scientific community has discarded its belief in matter as the fundamental building block of the universe as an antiquated relic of a mistaken scientific past. This being the case, it is likely that our societal obsession with "stuff" will soon join it on the scrap heap of history. In the future, the real power brokers will be those who understand and know how to manage the exchange of energy between the visible and invisible worlds.

A century of discovery in the field of quantum physics has reaffirmed the essential truths left to us by our spiritual ancestors. For those who know how to see, there is magic everywhere. Things that appear solid, fixed and unchanging are really just momentary snapshots of flowing currents of energy. Energy responds to coherent consciousness, shaping itself to fulfill the expectations and beliefs of the observer. Flowing energy brings new life with each and every breath. Energy that doesn't flow stagnates. Your emotional state establishes the flow rate of energy in your life and helps shape the world around you.

A heartfelt understanding of simple principles like these allowed Jesus to heal the sick, raise the dead and walk on water. It explains how the energetic signature of a substance can remain in a homeopathic solution so dilute that none of the original substance remains. It explains how Reiki Masters can manage healing energy at a distance. It explains the placebo effect — and the cases of spontaneous remission we doctors have always found so mysterious. It explains why great works of art and music continue to appreciate in value across the centuries, while last year's high-end tech toys are already piling up in landfills.

Your elementary school science teacher taught you that energy can neither be created

*Mother Teresa (1910 – 1997)*

The more you have, the more you are occupied, the less you give. But the less you have, the more free you are. Poverty for us is a freedom.

– Mother Teresa

nor destroyed — but it can change forms. She was right. She was talking about immortality. She was talking about you.

Learning to walk in that state of innocent grace is humanity's legacy and birthright. It is your legacy and birthright. It is the channel through which life flows.

You have the power to change your life. You have the power to change your destiny. You have the power to change your world. It doesn't matter whether you are living in a mansion, an inner-city tenement or a nursing home. It doesn't matter whether you are rich or poor, in radiant health or struggling to overcome a serious illness.

You have the power to change the world.

The change begins with you.

EPILOGUE

*Spotted Wolve Spirit, a leader of the Lakota Sweat Lodge Ceremony*

## YOU ARE CORDIALLY INVITED

There is great wisdom to be found in Native American spiritual traditions. The Vision Quest—a time set aside for the solitary, disciplined exploration of one's personal connection with the spirit realm—silences the noise and chatter of everyday life long enough for us to hear the voice of sacred wisdom. The Vision Quest offers a special opportunity to seek guidance, purpose and direction during challenging times. Those engaged in the Vision Quest gain access to the spiritual and energetic resources that will help them move forward in life.

The Vision Quest is an intensely personal process. No one can do it for you. The responsibility for seeking purpose, destiny and direction in your life rests on your shoulders alone. The solitary nature of the Vision Quest honors the fact that your connection to the infinite wisdom and limitless resources of the quantum realm is a direct line. Only you can access it.

For many, a serious illness can become a kind of Vision Quest. Illness calls us away from the ordinary and allows us to view life from a new perspective. Suffering is a lens that focuses our attention on things that matter with a clarity nothing else can match. Although you may have the love and support of others, the journey of illness is ultimately one we each make alone. Like the Vision Quest, the solitary nature of illness teaches us to rely on our own inner resources and renew our personal connection with the source of life. When we recover, we return to society wiser and stronger for having gone through the experience.

I have had the honor and privilege of

participating in many Native American Sweat Lodge Ceremonies, during which I have witnessed dozens of life-changing health transformations among participants. This sacred ritual is another time of reflection, sanctification, purification and refocusing on things that matter, but unlike the solitary Vision Quest, the Sweat Lodge Ceremony is a communal event held under the guidance of a respected elder. Participants in the Sweat Lodge Ceremony do not pray for themselves—they pray only for others. There is wisdom in this too.

While the Vision Quest develops our individual connection with the world of spirit, the Sweat Lodge Ceremony reminds us that we are all part of each other. Praying for another is an act of love and generosity that expands our energetic bandwidth and opens us to an enhanced flow of life-giving energy. This is especially important during times of illness, when praying for our own needs would keep our attention focused on our fears, doubts and misery—the very emotional states that hinder our ability to recover.

The easiest way to avoid this kind of emotional quicksand is to invest your prayers in someone who needs them more than you do. I can assure you they're out there, and your efforts will make an enormous difference in their lives and yours. Being a conduit for healing in the lives of others is a simple and effective way to heal yourself.

It would be difficult to overstate the importance of community as a factor in healing. Dr. Dean Ornish's well-researched book *Love and Survival*[7] shows that love and intimacy can alter the course of disease, disability and survival by as much as five hundred percent! Nothing in modern medicine can match this level of safety, efficacy and affordability.

Given such amazing clinical results, you'd think love and caring would be top priorities at every level of the health care system. Unfortunately, that's not the case. America's health care system was once the best in the world, its success attributable to a direct,

---

[7]Ornish, Dean, *Love and Survival* (New York, HarperCollins Publishers Inc., 1998)

caring partnership between patient and physician. Long-term relationships between patients and physicians promoted trust and a deep understanding of the role lifestyle and relationships play in illness. Because doctors instinctively understood that a patient's illness represented more than just an immediate set of symptoms, they could offer something more effective than a quick prescription. Patients who required hospitalization recovered in quiet surroundings with the help of a compassionate nursing staff whose services routinely included a nightly back rub.

Things have changed. Physicians and patients now have little say in establishing an appropriate standard of care or choosing an effective strategy for recovery. Those decisions are now being made by insurance company administrators who've never even met any of the parties involved. Those decisions are not based on the patient's best interests — they're focused on the insurance company's bottom line.

As a result, medicine has been depersonalized and continuity of care has become a thing of the past. Employer remarketing of insurance plans leaves patients scrambling to find new physicians every few years. Physicians taking on new patients must try to assimilate a lifetime of medical history, assess current symptoms and prescribe a course of treatment in the few minutes allotted by insurance companies for a typical office call. Rates of hospital-acquired infection and iatrogenic fatalities are embarrassingly high. The morale of countless dedicated health care professionals who are trying to make the best of a failing system is at an all-time low. Doctors and nurses alike are leaving the profession in droves. The back rubs are long gone, and today's hospitals sound more like video arcades than temples of healing.

On June 21, 2000, the World Health Organization released the findings of a comprehensive study of the overall quality of health care around the globe. One hundred and ninety-nine nations were assessed in the study. Despite ranking first in per capita spending, the United States ranked thirty-seventh in overall quality of health care. When judged by infant mortality rate—a scale widely used to determine overall levels of health care access—the U.S. ranked twenty-eighth. In terms of financial fairness, the U.S. ranked fifty-fourth. Forty-five nations ranked higher than the U.S. in terms of life expectancy.

America's health care system was once the finest in the world. What happened?

In a nutshell, bureaucracy! Medical care in the United States is not customer driven. Fifty-five percent of the $1.6 trillion spent each year on U.S. health care is funneled through government-run Medicare and Medicaid systems. Since government controls the majority of the U.S. health care market, its priorities set the standard of care for the entire industry. As a result, high-quality, personalized care has been deprioritized in favor of the presumed "efficiency" of the industrial model of mass production. The one-size-fits-all approach to medicine that has resulted from this philosophical sea change has never worked well in the clothing industry. In health care, it's been a disaster.

The system's top priority is no longer the patient—it's preserving the system itself. We are now stuck with a costly and inefficient system that relies heavily on standardization to function. Neither patients nor health care providers are happy with things as they are, but the system's heavy reliance on standardization makes change and innovation from within nearly impossible.

Things are about to get worse. Medicare's legendary bureaucracy, inefficiency and dismal repayment rates have forced many physicians out of the system. No matter how compassionate your doctor may be, she is losing money on her Medicare patients. Taking on too many of them will bankrupt any practice. This simple economic reality is already limiting many seniors' access to health care. Unless substantial sources of new funding appear, the millions of baby boomers now approaching retirement age will crash the system entirely.

Frankly, it's hard to imagine where those new revenues might come from. Skyrocketing costs have led an increasing number of employers to drop health coverage from their benefits packages. This has left millions of hardworking Americans and their children uninsured. If employers and employees can't afford coverage for themselves, who's left to rescue Medicare?

It can be difficult to recognize the early signs of a revolution—especially when you're in one. Revolutions begin when inequity and widespread public dissatisfaction spark demands for change. Those who've risen to the top of the existing system resist change because they benefit from keeping things as they are. Their efforts to shore up the current system may keep the lid on the pot for a while, but if things don't change, the pressure will continue to build until the whole thing blows.

In the political arena this kind of revolution often leads to tragedy. It does in the health care arena, too, but the signs are harder to recognize. When a health care system breaks down, our sleep isn't disturbed by the sound of gunshots and bodies don't pile up in the streets. When health care fails, people suffer quietly in homes and hospitals until they're tucked away in morgues.

Fortunately, you don't have to be one of them. None of us do. Believe it or not, things are not as bleak as they might seem. When a tree falls in the forest, the nourishment it has received from soil and sunlight is released back into the earth through a process of decay, and radiant new life soon springs up all around it. The old health care system may be dying, but wellness itself is thriving—if you know where to look.

The Wellness Revolution has begun. Legions of ordinary men and women are rising up and learning to care for themselves and one another in new ways. Physicians and patients who've experienced the failures of the existing system are forging new alliances that extend beyond the boundaries of culture and conventionality. A marriage of the high-tech electronic medicine of the West and the high-touch energy wisdom of the East has birthed a generation of technologies with staggering potential to shape your future—and the future of medicine.

We are learning from one another. We are sharing with one another. We are exploring new terrain in healing, wellness and the art of being human. This movement toward love, wholeness and inner peace is being driven by a heightened state of human consciousness. I've spoken to audiences around the world; the change is palpable everywhere. A strong sense of unity is emerging that transcends racial, religious and national boundaries.

Human beings instinctively grasp quantum concepts because these principles harmonize with their own life experiences. The confidence that comes from knowing that the hard facts of science align with the teachings of the great spiritual masters and our own heartfelt perceptions fosters the kind of intellectual and emotional coherence that can change the world.

I find tremendous hope and reassurance in knowing that so many ordinary human beings around the world are experiencing this transformation together. Grandmothers in Tokyo, shopkeepers in Tel Aviv and students in Tehran are all moving toward a more inclusive form of wellness that honors the true nature of our interconnected universe. In finding their way to wholeness, they are finding their place in the whole. This bodes well for all of us. It's a sure sign that the future of mankind can be much brighter than some would have us believe.

If you've ever read a newspaper and wished things could be better, if you've ever looked in the mirror and wondered about your true purpose in life, if you've ever had a feeling that there is more to life than you've been told, then you are one of us. There is more, and your wondering is your soul's way of telling you that you are ready to find it. If this book has helped bring you to that state of awareness, then it has done its work—but our journey together has just begun.

This book includes two complimentary tickets to a new kind of healing event. There is no catch. They are a gift from my heart to yours. These tickets will grant you admission to a dynamic, interactive live seminar that will allow me to share the kind of practical, hands-on skills and information that can't be taught in a book.

You'll learn high-speed blues-busting techniques that can recharge your physical, emotional and energetic state in seconds. You'll discover revolutionary new tools that are shaping the future of medicine. You'll be introduced to unsung heroes working hard behind the scenes—and sometimes in exile—to establish new paradigms for health care. You'll take home a practical, step-by-step approach to wellness that can change the course of illness and transform your life.

More important, you will enjoy a day of hope, encouragement and community in the presence of wonderful, conscious people who are ready to change themselves and their world.

I hope you will join us.

For the many individuals who've attended our seminars and requested help in designing a personalized wellness and recovery plan, we've now created a series of three- and seven-day wellness retreats that will allow us to do just that. Hosted by a highly qualified team of enlightened health care providers in luxury spa locations throughout the world, these events are guaranteed to take you away from the ordinary and move you closer to your goal of radiant health. Comprehensive information

about the program is available online at www.drnoah.com.

The day I received official word of my early release from prison, my thoughts turned to Gary, the gentle man in the wheelchair with the championship smile whose visions I'd once dismissed as impossibly far-fetched. Gary's keenly honed intuition had allowed him to accurately predict my personal future and the future of world events. I wanted to thank him for his amazing insights and learn more about how he did it, but I couldn't. It was too late. Gary was gone.

Gary—a quadriplegic—had been one of my favorite patients for nearly twenty years. A self-taught webmaster, he'd learned to operate a computer with the aid of a simple homemade mouthpiece. He lived alone in a run-down mobile-home park but never complained about the lack of creature comforts offered by his meager existence. His kindness, perseverance and ability to overcome adversity were legendary. He was an inveterate punster who, if left to his own devices, would spend his entire office call using me as a practice audience for his latest round of jokes.

While I was away in prison, Gary died in a hospital from complications of a bed ulcer. His tragic death was completely preventable and a shameful indictment of our broken health care system.

For a quadriplegic, a broken wheelchair is a death sentence. When Gary's chair failed, he spent many months confined to his bed. I spent those months begging Medicaid to repair or replace it, but no action was taken. With my own financial resources exhausted by legal fees, there was little more I could do. Shortly after I left for prison, the inevitable happened. Gary developed multiple bed ulcers. The ulcers became infected. The infection led to the generalized septicemia that took his life.

Gary was a dear friend, a remarkable human being and a compassionate soul who will be missed by everyone who knew him. He deserved better. We all do.

If a million of us had each chipped in a penny, Gary would have had his wheelchair. If a million of us had each chipped in a prayer, he might not have needed it.

We live in a world of possibilities. We can make things better.

## ASKLEPIA INTERNATIONAL

Dr. Noah's proceeds from the Wellness at Warp Speed book, seminars and rejuvenation retreats will be donated to Asklepia International.

After a 2,500 year hiatus, Asklepius—the legendary god of medicine—is joining us to help spread love, peace, health and happiness around the world.

Asklepia International invites conscientious individuals and companies to help launch ASKLEPIA, a state of the art charity hospital cruise ship and wellness sanctuary.

Please join Dr. Noah in this international humanitarian campaign by contributing your time, expertise or financial support today.

To learn how you can make a difference in your world, visit us online at www.AsklepiaInternational.org or www.WellnessatWarpSpeed.com.

# BIOGRAPHY

Dr. Noah McKay graduated Magna Cum Laude from Tufts University with a Bachelor of Science in Biology, and in 1983 was awarded his Doctorate Degree in Medicine at the Albert Einstein College of Medicine in New York.

He personally encountered the limitations of his Western medical training while hospitalized with heart failure at the age of thirty-three. When drugs and surgery offered no hope, he reached beyond the box of conventional medicine and found his cure in the subatomic space of his own body—a hidden world mapped by scientists working for more than a hundred years in the field of quantum physics.

These exciting new discoveries helped him to establish the largest private integrative medical practice in the state of Washington. Dr. Noah's pioneering efforts to make innovative medical care available to the general public were highly popular with over 30,000 patients, but met with stiff resistance from regulators in the insurance industry and the U.S. Department of Justice.

To further the cause of freedom and choice in health care, in 2001 he willingly served a year in a U.S. Federal Prison Camp. His struggle helped bring about important changes in the insurance industry that now allow patients to receive coverage for complementary and alternative care. More importantly, the sanctuary of prison introduced Dr. Noah to Love—the world's most powerful healing technology.

In 2003 he underwent aortic valve replacement surgery with an Omni 3000 prosthetic valve.

Today Dr. McKay shares his transformational message of love, hope and practical quantum healing techniques with enthusiastic audiences around the world.

*Wellness at Warp Speed*

# RECOMMENDED READING

The following writings will deepen your understanding and appreciation of quantum principles we've only touched on in this book. Learn to apply these principles in your own daily encounters and watch the magic unfold!

### Autobiography of a Yogi
by Paramahansa Yogananda

You won't fully appreciate the significance of this book until you understand how quantum phenomena really work. Yogananda lived in the modern world as a shining example of a quantum human being.

### Quantum Healing: Exploring the Frontiers of Mind/Body Medicine
by Dr. Deepak Chopra, M.D.

Contemporary scientist, poet, philosopher and author of more than forty books, Dr. Chopra is a master of conscious living and the new science of quantum spirituality. I highly recommend all of his writings, but this volume offered the greatest help in my time of need.

### The Holographic Universe
by Michael Talbot

This is the book that introduced me to the fascinating world of quantum mechanics back in 1991. This classic work is a wonderful primer for those just beginning their exploration of quantum science.

### Tao of Physics
by Fritjof Capra, Ph.D.

Dr. Capra draws beautiful parallels between quantum physics and the traditional spiritual teachings of Buddhism and Hinduism.

### The Dancing Wu Li Masters: An Overview of the New Physics
by Gary Zukav

A poetic and highly readable introduction to the scientific and philosophical implications of quantum physics.

### Love & Survival: The Scientific Basis for the Healing Power of Intimacy
by Dean Ornish, M.D.

Dr. Ornish offers convincing scientific evidence supporting the claim that love is the world's most powerful healing force.

### Jump Time: Shaping Your Future in a World of Radical Change
by Jean Houston, Ph.D.

Dr. Houston is one of the most brilliant scholars, philosophers, authors and visionary thinkers of our time. In *Jump Time*, she explores the potential of new global paradigms and the rebirth of human society.

## Chicken Soup for the Soul
*by Jack Canfield and Mark Victor Hansen*

No one is better known and more respected in the field of human potential than Mark Victor Hansen. This heartwarming collection of stories, told with the simple mastery of an Aesop, will help reshape your personal vision of the possible.

## The Art of War
*by Sun Tsu*

This 2500 year old collection of essays is revered throughout the world for its perennial applicability to the human condition. It offers guidance in the art of relationships, negotiations and everyday human entanglements.

## The Mozart Effect
*by Don Campbell*

Campbell shares revolutionary discoveries in the study of sound and makes a compelling case for the therapeutic effectiveness of music.

## The Quantum Doctor
*by Amit Goswami, Ph.D*

Dr. Goswami, featured in the movie "What the Bleep Do We Know," explores the intersection of quantum physics, integral medicine and human consciousness in a precise and easy to understand manner.

## Physics and Reality, The Revolution in Modern Science
*by Werner Heisenberg, Ph.D*

A Nobel Prize-winning physicist and founding pioneer of quantum mechanics explores the impact of science on culture through the mechanism of philosophy.

## Power vs. Force: The Hidden Determinants of Human Behavior
*by Dr. David R. Hawkins, M.D., Ph.D.*

The Carl Jung of our generation, Dr. Hawkins assigns human emotions an energetic frequency value and assesses the impact and effectiveness of human behaviors, motivations and attitudes. We can use his frequency charts to monitor our progress towards optimal health.

## Your Body's Many Cries for Water: You Are Not Sick, You Are Thirsty
*by Fereydoon Batmanghelidj, M.D.*

Author and researcher Dr. Batmanghelidj—the Water Guru—has written more than seven books tracing the etiology of heart disease, cancer, stroke and diabetes to chronic and sustained dehydration.

### Cell Phones: Invisible Hazards in the Wireless Age
by George Louis Carlo and Martin Schram

$28.5 million dollars of research conducted over a period of six years convinced scientist, epidemiologist and attorney Dr. George Carlo that EMF radiation is the greatest health epidemic facing the planet.

### The Hidden Messages in Water
by Dr. Masaru Emoto and David A. Thayne

Dr. Emoto's fascinating pictures of water crystals reveal the physical side of invisible quantum molecular interactions.

### The Power of Now: A Guide to Spiritual Enlightenment
by Eckart Tole

Living in the present moment is natural for children, but it is learned behavior for adults. Tolle offers strategies to help grown-ups find their way back to enlightenment.

### Vibrational Medicine
by Richard Gerber, M.D.

Dr. Gerber's excellent book is arguably the best summary of advancements in the emerging field of energy medicine to date.

### Wholeness and the Implicate Order
by David Bohm, Ph.D.

A classic work on the indivisible nature of the universe written by a legendary pioneer in the fields of quantum mechanics, philosophy, neurophysiology and higher consciousness.

### The Elegant Universe
by Brian Greene, Ph.D.

Physicists have been arguing about the implications and interpretation of their discoveries for nearly a century. Greene offers a highly readable introduction to the debate and explores superstring theory as a way out of the dilemma.

## ACKNOWLEDGMENTS

I would like to start by thanking my incredible wife Kim and our wonderful children Zachary and Zaira for their never ending love and kindness. I also want to thank my mother Mary, my brother Nader and my sister Nini for their prayers and constant vigilance. I am especially grateful to Myrna and Jay Sapunar, who graciously abandoned a comfortable retirement lifestyle in sunny California and moved to Oregon to look after my family during my imprisonment. Thank you, all of you. Your love and strength have helped me keep my commitment and focus on the higher missions in life.

My deepest appreciation goes to the Malek family, to Shoky for believing in me, and to Moosa and Shohreh for generously funding the publication and international distribution of this book. Your hope, courage and vision will touch more lives than you can imagine.

Heartfelt thanks to the hundreds of courageous friends and patients who stood by me through an unimaginably difficult period of my life. You know who you are. Whenever I was in need of love, encouragement, advice, hospitality or financial support, you were there.

Words will never adequately convey the depth of my gratitude.

The seeds of this book were planted in a prison. I would like to thank the inmates I served time with for sharing the wisdom, courage and inspiration that helped it take root and grow. I would also like to thank the guards and administrators of the U.S. Bureau of Prisons and Department of Justice for giving me the time and deep solitude I needed to restore peace and wellness to my life.

Special thanks to my good friend Cathy Tomlan, who has worked countless hours painstakingly researching and editing my words and ideas. Her work on the book has been a labor of love; it could not have been published without her. Many thanks, also, to the wonderful staff at Palace Press for contributing to the book's beauty and visual appeal. I especially want to recognize the gracious and talented Barbara Genetin for her patient endurance and moving artistic renditions. Stand up and take a bow, Barbara. You deserve it.

To all my good friends around the world, please know that you are the source of my passion and enthusiasm for life. I love you, I thank you, and I hope our adventures will bring us together again soon.

謝

arigato

تشكرم

Thank you

gracias

யுப்பௌ

merci

σαςευχαριστούμε

# COLOPHON

**Publisher & Creative Director:** Raoul Goff
**Executive Directors:** Peter Beren & Michael Madden
**Art Director:** Iain R. Morris
**Designer:** Barbara Genetin
**Acquisitions Editor:** Mikayla Butchart

The original paintings on pages 26, 30, 32,
42, 70, 74, 80, 119, 123, and 136 were
created by Noah McKay.
**Managing Editor:** Cathy L. Tomlan

Back cover photo courtesy
of Karen Orders Photography
http://www.karenorders.com

Mandala Publishing would also like
to give a very special thank
these people for their additional
design and editorial support:
Jennifer Gennari, Valerie Reckert,
Sonia Vallabh and Carina Cha.